EMD F-UNIT LOCOMOTIVES

BRIAN SOLOMON

COPYRIGHT © 2005 BY BRIAN SOLOMON

Published by
Specialty Press Publishers and Wholesalers
39966 Grand Avenue
North Branch, MN 55056
United States of America
(800) 895-4585 or (651) 277-1400
http://www.specialtypress.com

Distributed in the UK and Europe by
Midland Publishing
4 Watling Drive
Hinckley LE10 3EY, England
Tel: 01455 254 450 Fax: 01455 233 737
http://www.midlandcountiessuperstore.com

ISBN-13 978-1-58007-083-6
ISBN-10 1-58007-083-3

All rights reserved. No part of this book may be reproduced or transmitted in any form or by any means, electronic or mechanical, including photocopying, recording, or by any information storage and retrieval system, without permission from the publisher in writing.

Material contained in this book is intended for historical and entertainment value only, and is not to be construed as usable for locomotive or component restoration, maintenance, or use.

Printed in China

Front Cover: *On September 19, 1990, former Alaska Railroad FP7s Nos. 1510 and 1512 were at work on the Wyoming Colorado Railroad at Walled, Colorado. Note the extra large winterization hatches. (photo by Brian Jennison)*

Title Page: *VIA Rail's Super Continental at Geikie, near Jasper, Alberta, is led by Canadian National FP9 6513 plus two more Fs. (photo by Tom Carver)*

Back cover, top left: *Burlington Northern No. 834, former Northern Pacific No. 7009D, is an F9. Unmodified F9s can be distinguished from earlier models because of the additional louvers on each side ahead of the first porthole window. (photo by Tom Carver)*

Back cover, bottom left: *On September 3, 1981, Rio Grande No. 5771 leads the Rio Grande Zephyr at Denver, Colorado. This Rio Grande F9A and F9B No. 5762 are now preserved at the Colorado Railroad Museum in Golden, Colorado. (photo by Dave Rector, Brian Jennison collection)*

Back cover, top right: *Engineer's controls on a Bangor & Aroostook F3A: Top left is the 8-position throttle, middle left is the reverser, bottom left is motor transition, on the right is the airbrake control. The semi-circular gauge in the middle is the traction motor load meter. (photo by Brian Jennison)*

TABLE OF CONTENTS

Introduction 4

Acknowledgments 5

Chapter One – Electro-Motive FT 6

Chapter Two – Postwar Fs 28

Chapter Three – New Haven FL9 76

Chapter Four – Recycled F Units 88

Bibliography 111

INTRODUCTION

Locomotives are naturally one of the most fascinating aspects of railroading. They are power. In the 1940s and 1950s, American railroads made the historic transition from steam to diesel-electric power. This was a fascinating and tumultuous time in railroading. The diesel forever changed how railroads operated. General Motors was the largest force driving dieselization. Its two largest competitors were Alco and Baldwin, which had been the two largest steam locomotive manufacturers. By the early 1950s, GM prevailed and dominated the locomotive market. At this crucial juncture its best-selling models were the four-motor F units. We remember the F unit today not just because it was a great locomotive, but because it looked and sounded great. At that time streamlining was in vogue, and the F was the most numerous streamlined locomotive of all time. Electro-Motive's 567 diesel primemover was tops in its class, and its sounds reverberated all across the American landscape.

At one time or another, F units have operated virtually everywhere on the North American railroad network. They have hauled heavy freight, pulled luxurious passenger trains, drilled yards, worked branch-line mixed trains, and delivered the daily commuters. The image of the F-unit's "bulldog nose" is a universal image of railroading. It has appeared on the cover of innumerable books and magazines, in generations of railroad literature, on coffee mugs, T-shirts, on toy and model trains, and even on graffiti in alleyways in Prague, Czech Republic. For many people, the F-unit symbolizes American diesel traction. The F-unit was the standard diesel on many American railroads for over two decades. Today, nearly 45 years since the last F-unit rolled out of Electro-Motive's La Grange, Illinois, plant, a few are still working in daily service. In 1997, *CTC Board Railroads Illustrated* estimated there were 314 existing F-units from a total production of 7,612. Some of these have since been scrapped. Many are preserved in museums and on tourist railways. When the last F turns a revenue mile, its place in history is ensured.

The Electro-Motive Corporation was founded in 1922, and during the 1920s it became the leading seller of self-propelled gas-electric railcars. These assorted machines were popular with many railroads for light passenger service on branch lines, but more importantly they represented an evolutionary step in the development of the modern diesel-electric locomotive. Interestingly, EMC didn't actually build railcars but rather coordinated their construction. Its primary gas-engine supplier was the Winton Engine Company. The market for gas-electric railcars dropped off sharply with the onset of the Great Depression in 1929.

In 1930, automotive giant General Motors bought both the Winton Engine Company and EMC. During the 1930s, subsidiaries of GM, EMC, and Winton revolutionized the way railroads considered motive power. In 1933, Winton introduced its 201A engine, a compact, high-output, two-cycle diesel. Then in 1934, EMC debuted America's first high-speed diesel-electric streamlined passenger train, Burlington's famous *Zephyr*, powered by the 201A. During the next few years, EMC built dozens of lightweight streamlined passenger trains for American railroads. In 1936, it also began producing diesel-electric switchers. The development of its streamlined E-units, which used six-wheel trucks in an A1A-A1A wheel arrangement, was a logical outgrowth of EMC's early streamliners. The E-units used a pair of diesel engines inside a styled truss carbody with an elevated cab and reinforced nose for crew protection.

In the 1930s, railroads still relied on steam power for most service, and diesels remained a novelty. This view changed in 1939, when EMC introduced its FT diesel, designed not specifically for passenger service but rather to move heavy freight trains. With 5,400 hp output it was intended to rival the most powerful steam locomotives of the time. This machine made the diesel-electric locomotive a practical heavy freight hauler, paving the way for total transformation of American railroad motive power. Within a decade of its launch, the production of commercially built steam locomotives had concluded. In little more than two decades, the American steam locomotive had been entirely supplanted.

Shortly after EMC debuted its FT, General Motors consolidated its control over both Winton and EMC. On January 1, 1941, EMC became the Electro-Motive Division of General Motors. It has become common to refer to GM diesels as EMD products. In this book, to avoid confusion stemming from the corporate consolidation, "Electro-Motive" is used to describe GM's diesel builder in the United States. This covers both the EMC and EMD eras.

Some images in this book may be familiar to students of the Electro-Motive diesels, but many are previously unpublished. I have scoured through thousands of photographs searching for interesting and distinctive images of F-units on North American railroads. In addition, I've made hundreds of my own F-unit photographs over the years, a few of which are presented here. My intent was to show the variety of the railroads' F-units, using diverse images to depict the machinery, the heterogeneous paint liveries, and also to show the services they performed and the areas they worked. Collectively, these images should convey the F's machinery and workings with the spirit of the times and places they have labored. While many different F-unit operators are covered, I've made no attempt to show each and every owner, nor represent every known livery. Such an effort is beyond the scope of this modest publication. I have chosen to feature several railroads with variegated images — Boston & Maine's FT fleet, Southern Pacific's F7s, and New Haven's FL9s are given more extensive photographic coverage than other lines. I've also chosen to cover several specific locomotives in great detail, photographing them inside and out. Former Alaska Railroad No. 1508 gets the best coverage. This locomotive is one of a few F7As that remains active and therefore may be of special interest to railroad historians. As of this writing, the locomotive resides in Utica, New York, and operates seasonally on the Adirondack Scenic Railroad.

Electro-Motive's F units have long been one of the most popular locomotives. They have spent many years of hard work, hauling freight and passengers for American railroads. But their image goes well beyond the real-life locomotives. Tens of thousands of children have enjoyed model Fs, myself included. It is my hope that this book furthers your interest in these great machines and helps you better understand why they came about, how and where they work, and the technology that made them special. Enjoy!

Acknowledgments

This book tells the story of Electro-Motive's famous F units, certainly one of the most recognizable locomotives in American history. Unfortunately, I cannot convey the sounds of F-units at work. Imagine, if you will, four 16-567 diesels working together up grade with a heavy train in tow. Their traction motors whine, exhaust stacks bark, and the ground shakes as they pass. When these engines work in multiple, they produce an unmistakable resonance that's difficult to describe, yet as much part of the F-unit experience as pictures and words.

In the process of assembling this book I was assisted along the way by many people. Without their help, what you see before you would not have been possible. I am not as much an author as a conduit for information. First I would like to thank the many photographers who graciously lent photos for this project. Each is credited by their published work. Special thanks to Jim Shaughnessy for providing many rare historic photos of F-units in action, also to George C. Corey, Stanwood K. Bolton, Robert A. Buck, Patrick Yough, George Kowanski, J. J. Grumblatt Jr., Tim Doherty, J.R. Quinn, Pete Ruesch, John Gruber, Jay Williams, Tom Carver, Don Marson, Brian Jennison, Bob Morris, and Fred Matthews. My father — Richard Jay Solomon — provided photographs, first brought me to see an F-unit at age 3 or 4, and among other things, helped proofread this text. Special thanks to Doug Eisele, who gave me unlimited access to photographs in his collection and access to his library.

Researching a project such as this involves many hours searching for information, scouring libraries, reading books, magazines, manuals, and timetables, not to mention discussions and interviews with railroaders and locomotive professionals. The Irish Railway Record Society in Dublin provided unrestricted access to their library. John Gruber helped with research and has provided a short sidebar. Pat Yough, Tim Doherty, and my father also lent me books and other information. Special thanks to Tom Carver, not only for providing valuable Electro-Motive literature, but also for organizing a first-class F-unit experience on the Adirondack Scenic Railroad. David Swirk and the employees of the Pioneer Valley were also extremely kind and helpful. David spent many hours answering questions on 567 engines and the operation and maintenance of CF7 diesels. Ten years ago Mike Danneman, Tom Danneman, and I made a pilgrimage to the Minnesota Iron Range to photograph the Erie Mining F9s. Without them, I could not have made the photos nor experienced A-B-B-A Fs in action. Thanks to T. S. Hoover for trips to photograph F-units on MBTA and Metro-North. Also, thanks to F.L. Becht of the Louisiana & Delta who provided access and a cab ride on CF7s in Louisiana, and the late John Conn for helping me with the Mass-Central. Mike Gardner lent the use of his darkroom and studio. Thanks to Dick Gruber, the National Railroad Museum at Green Bay, Wisconsin, the Monticello Railway Museum in Illinois, my brother Seán, my mother Maureen, and to Tessa Bold.

CHAPTER ONE

ELECTRO-MOTIVE FT

Electro-Motive 103, the four-unit FT demonstrator in resplendent dark green and cream paint, works eastward under wire on the Boston & Maine at North Adams, Massachusetts, in 1940. (photo by Victor Newton, Robert A. Buck collection)

Electro-Motive 103

In November 1939, Electro-Motive quietly released its latest and most significant product, four-unit road diesel No. 103, the demonstrator FT model. This handsome streamlined machine, dressed in dark green and cream with "GM" on its nose and "Electro-Motive" on its sides, looks familiar to us today. It's hard to believe that in its day it was a radical new machine. Electro-Motive 103 was the pioneer road-freight locomotive designed for mass production and represented the culmination of 10 years of research, development, and carefully crafted market strategy. Today, production-built diesel-electric locomotives haul virtually all rail-freight in North America, but in 1939 that was not the case. Steam ruled American rails for over 100 years. On the eve of World War II steam power did most of the work, although a few railroads were using electric locomotives. Diesels were just minor players, assigned as yard switchers and on a handful of high-profile lightweight streamlined passenger trains.

Electro-Motive's FT would soon change the way railroads viewed diesel-electric power. Not only would it demonstrate that diesel-electrics could be used as practical road-freight power, but that they were in fact vastly superior to conventional steam power.

Despite FT's extraordinary significance to American railroad history, it debuted quietly. It was not accompanied by the fanfare that one might expect. Unlike earlier Electro-Motive pioneers, such as Burlington's *Zephyr*, which had been closely followed by the media, the FT toured the United States largely unnoticed for months. Rather than boast, General Motors let the FT tell its own story.

In February 1960, TRAINS Magazine's David P. Morgan explained in his article, "The Diesel That Did It," that in 11 months No. 103 traveled 83,764 miles on 20 Class I railroads in 35 states. He noted that the FT was the locomotive with the single greatest impact on U.S. railroading. In effect, it was the FT that convinced many railroads to dieselize. The FT demonstrated that Electro-Motive could mass-produce road diesels matching steam locomotives in demanding freight service. A pair of FT units equaled the output of the typical 2-8-2 Mikado or 2-10-2 Santa Fe, two of the standard "workhorse" steam types, while a four-unit FT set could rival even the most advanced "Super-powered" steam. Most impressive was the FT's extraordinary high starting tractive effort. Unlike reciprocating steam, a diesel-electric can produce maximum tractive effort from the start, giving the FT an inherent operating advantage. Another advantage of the FT design was its B-B wheel arrangement. Each unit placed all of its adhesive weight on driving wheels. The FT's enormous starting tractive effort rivaled that of heavy electric locomotives. The FT had most of the advantages of an electric locomotive without the need for costly overhead electrification.

Electro-Motive 103 demonstrated its exceptional capabilities in a variety of services and consistently proved that its

Above: The Missouri Pacific's first locomotive, built in Massachusetts and shipped via sailing vessel and steamboat to St. Louis in 1852. ★ *Below:* One of the several 5400-horsepower Diesel-electric locomotives now hauling freight trains over the Missouri Pacific Lines.

General Motors sold hundreds of mass-produced FTs between 1940 and 1945. Missouri Pacific was one of several large railroads to take advantage of these powerful freight diesels. (author's collection)

new diesel could out-haul modern steam in the most demanding environments. Morgan explained that the four-unit set made an impressive showing on Baltimore & Ohio's famous Cranberry Grade that climbs eastward from M&K Junction near Rowlesburg, West Virginia, to the summit at Terra Alta. Cranberry Grade was among the steepest double-track mainlines in the United States. Long the bane of railroaders, it typically required multiple steam locomotives fore and aft to move tonnage freight. It was here that No. 103 lifted 1,952 adjusted tons in 41.3 minutes, averaging 16.6 mph without a helper.

In another example of the FT's power, Morgan described its performance on Southern Pacific's famous Tehachapi crossing. Santa Fe had trackage rights on this tortuous single-track mountain mainline to reach northern California, and trains of both railroads shared the titanic struggle over the mountains, which required navigating the legendary Tehachapi loop. Morgan explained that one of Santa Fe's modern 2-10-4 Texas-type locomotives could move 1,100 tons in 2 hours and 15 minutes between Caliente and Tehachapi Summit on the west slope of the mountain crossing. Likewise, one of SP massive cab-ahead articulated 4-8-8-2 steam locomotives would move 1,350 tons in the same time frame. Both locomotives were state-of-the-art Superpowered steam. On the same grade, 103 hauled an estimated 1,800 tons in just 1 hour and 31 minutes. Unlike steam, 103 did not need to pause for water stops. This made it especially attractive for desert operations where water is scarce.

High Tractive Effort

The 103 was geared for 75 mph operation, and each unit had a maximum output of 1,350 hp, giving the four-unit set a 5,400-hp rating. Each unit delivered approximately 32,500 lbs. tractive effort, giving the set a total of roughly 228,000 lbs. tractive effort. The FT set weighed roughly 912,000 lbs. High tractive effort gave the FT a great hauling advantage over all contemporary steam power. According to Alfred Bruce in his book *The Steam Locomotive in America*, Santa Fe's 2-10-4 weighed 908,000 lbs. (including the tender) and, based on a factor of adhesion of 4.1, delivered 93,000 lbs. tractive effort. In other words, the FT delivered more than twice the amount of starting tractive effort for the same weight. In April 1941, *Railway Mechanical Engineer* compared a production four-unit FT set (weighing 923,000 lbs.) with three of the most powerful steam locomotive types ever built — Northern Pacific's 2-8-8-4 *Yellowstone*, Virginian's massive 2-10-10-2 Mallet compound, and the legendary Erie *Triplex* that used a 2-8-8-8-2 wheel arrangement. The tractive effort delivered by these machines was respectively 153,300 lbs. (with booster engine), 176,600 lbs. (working as a compound), and 160,000 lbs. (working as a compound). The point of this exercise was to prove that the FT far exceeded the tractive effort of the most powerful steam locomotives ever built, not to mention the common 2-8-2 Mikado that most railroads were using to haul freight.

Electro-Motive demonstrated that the FT could start a heavier train than a steam-powered locomotive. It logically followed that in day-to-day operations, using road diesels could allow the operation of longer and heavier trains, saving money. Their greater power could also eliminate some helper districts.

Carbody Design

Like Electro-Motive's passenger "E-units," the FT employed a streamlined carbody. The construction of the carbody was more than just a decorative covering; it was integral to the structure of the locomotive. Carbody framing was like that of truss bridge, and it was constructed of welded carbon-molybdenum steel fastened to the locomotive platform. This was made from steel bolsters and covered heavy steel plate. Evidence of the carbody truss-framing is seen through the intake vents and porthole windows on the sides of the FT units. Covering the framing were side panels made from 3/8-inch–thick plywood sheets, sandwiched with sheet steel. Panels were clamped in position to avoid the use of bolts or rivets on the side panels, thus allowing for a smoother, streamlined appearance.

A proud railroader, presumably the locomotive's engineer, poses with near-new Santa Fe No. 102, one of the first production FT sets. Santa Fe was the first railroad to make significant use of road freight diesels. (author's collection, photographer unknown)

In their original design, the FT's were intended to run either in four-unit or two-unit combinations. The cab units are called "A" units, while the cab-less units are known "boosters," or "B" units. Initially these units were viewed as locomotive sections. It was typical to describe a four-unit (an "A-B-B-A" set) FT as a single 5,400-hp locomotive. As originally intended, each A-B pair was semi-permanently coupled together. They were attached directly with a drawbar connection instead of a common coupler and shared some electrical components. These A-B pairs were like Siamese twins that shared a few common organs — they could not function independently of one another.

If the individual "units" were viewed as independent machines, railroads and Electro-Motive feared that railway labor would fight adoption of diesel-electric locomotives or make impractical demands that each "locomotive" should be manned. From labor's viewpoint, manning each unit was consistent with steam locomotive practice in which each and every locomotive needed both an engineer and a fireman. Eventually, American railroads negotiated with labor, and individual crews were not required on each unit, but railroads agreed to retain the fireman's position on diesels, despite the obvious lack of a fire to tend. However, while a fireman would no longer need to tend to the locomotive boiler, he was still required as an extra set of eyes on board the locomotive and was also needed to care to the needs of the diesel en route. In addition to firemen, some railroads assigned traveling mechanics to FTs in the early years.

Carbody Construction

The FT's cab was elevated to provide crews a better forward view and a safer place to ride. The rounded nose was reinforced to provide crews with added protection in the event of a collision. Steam locomotives had inherently offered a degree of protection to operating men, since the mass of the boiler protected them in the event the locomotive struck something at speed.

The FT and subsequent F-unit models featured front-end plows as well as anticlimbers, which were located above the plow at frame-level. Exterior cab styling took aesthetic cues from GM automobile design. The cab had two forward-facing windows, which echoed automobile practice. The style, slope, and angle of the windshields closely resemble that of 1939 Buick models. Inside the cab, handles to roll down windows were nearly the same as those used on GM automobiles. Entrance to the cab from the outside was afforded through a central nose door, and doors on both engineer's (right) and fireman's (left) sides of the locomotive. In addition, a centrally located door at the back of the cab provided access to the engine room. The nose and side doors opened inward. The engine-room door opened away from the cab. Seats were located on both sides of the cab and positioned to give the crew a comfortable forward

One of General Motors' World War II-era advertisements featured Milwaukee Road FTs. The Milwaukee initially used FTs to span the gap in its electrified territory between Avery, Idaho, and Othello, Washington. (author's collection)

view. Below the cab side doors, and additional side doors leading to the engine room, were ladders affixed to the carbody and situated directly over the trucks. Additional ladders were located at the very ends of the bodies.

The distinctive "bulldog nose" is comprised of four independent sheets of formed 12-gauge pressed steel and a cylindrical headlight housing welded to the locomotive carbody framing. Reinforced underframing provides structural support. The characteristic compound curves around the headlight housing were actually sculpted from automotive bonding putty. Traditionally, F-unit headlight assemblies consisted of a single-bulb unit in front of a one-piece reflector. Some railroads, such as Boston & Maine and Rio Grande, used multiple-bulb headlights in a hexagonal pattern.

To ensure crew safety, the F-unit frame had a fracture point built in behind the cab. This was designed to allow the locomotive body to collapse behind the cab in the event of a collision, providing safe area for the crew. An FT unit measured 14 ft. 1/2 in. tall, and 10 ft. 5-5/8 in. wide (accounting for handrails and other protrusions). A four-unit set was 193 ft. long, with an A unit measuring 48 ft. 3 in. FT A-unit trucks were centered 27 ft. 3 in. apart; on the B units, truck centers were 26 ft. 6 in.

The FT's most readily identifying external characteristics were its row of four evenly spaced porthole windows on each side of the locomotive, and two sets of recessed radiator fans on the roof to maintain a flush profile at each end of the carbody.

Air-intake vents were located along the sides near the top of the locomotive. Three large vents were on each side of the locomotive, with the center vent significantly larger than those to either side of it. Wire mesh and filters protected these vents. The center vents were for the engine room, and the side vents for the radiators located near the roof of the units, below the recessed fans.

The B units were built to accommodate a large steam generator, which was necessary for passenger operations. If used, the steam generator and water storage tanks were located at the rear end of the unit (the opposite end that coupled

Baltimore & Ohio was one of several eastern railroads to buy FTs. These diesels changed the way railroads moved freight. Within 10 years of their introduction, commercially built domestic steam-power production had ceased. (author's collection)

Electro-Motive's customers included the Atlantic Coast Line, as portrayed in this 1940s advertisement. (author's collection)

by drawbar to the A unit). As a result, from the side a FTB was asymmetrical — the boiler end had more than two feet greater overhang from the truck center than the drawbar end.

Carbody Preference

Carbody-style construction was the standard for most road diesels from late 1930 until the mid-1950s when hood-style road-switchers, became the dominant locomotive type. Although there were variations in the carbody between models, all Electro-Motive F units use the same basic construction. Electro-Motive "Fs" and other carbody locomotives are often described colloquially as "covered wagons," alluding to the wagons used by western pioneers. The analogy seems appropriate since the Fs were first used on a large scale by western lines such as the Santa Fe.

Since the 1950s, boxy utilitarian designs have predominated on American freight locomotives. Today, it may seem strange that railroads would order streamlined carbody diesels for freight service. There were good reasons for the carbody design when it was first introduced. In the 1930s, streamlining was in vogue. Everything from toasters and washing machines to new cars and passenger trains were designed with a streamlined aesthetic. In fact, the FT appears quite reserved compared to the more flamboyant designs applied to Electro-Motive's earlier streamlined passenger trains. Diesels were expected to look sleek and modern — streamlining was a tool to sell the locomotives to railroad officials, while also making them appealing and palatable to the public.

General Motors wanted its diesels to appear progressive, modern, clean, and colorful. The streamlined FTs were a sharp contrast to the tired and dirty old steam locomotives. Though marketed largely as a freight locomotive, the FT was designed to work in either freight or passenger service. It was expected that passenger locomotives look good.

The Electro-Motive FT demonstrators were sold to Southern Railway subsidiary Cincinnati, New Orleans & Texas Pacific in 1941. CNO&TP 6100A, pioneering FT cab unit 103A, was preserved at Museum of Transport in St. Louis. On September 9, 1972, the historic locomotive was displayed by Electro-Motive at McCook, Illinois. (photo by R.R. Richardson, Doug Eisele collection)

The streamlined carbody had practical concerns as well. Although diesels required less maintenance than steam locomotives and were designed to operate over hundreds of miles on a single run between servicing stops, it was common for mechanics to work on diesel-electrics en route. The reality of early diesel operations was that either railroaders or traveling Electro-Motive mechanics would routinely perform heavy locomotive maintenance, sometimes implementing complex repairs as they rolled along at speed. As a result, it was necessary to protect workers from the wind and elements, and this was handily accomplished by the protective carbody. End doors allowed passage from unit to unit or into a passenger train.

Inside the Electro-Motive FT

Aesthetics aside, it was the locomotive's guts that made Electro-Motive's FT a success. It was powered by Electro-Motive's latest and greatest diesel engine, what would become one of the most commercially successful diesel designs of the mid 20th-century — the famed 567 engine. Electro-Motive's early diesels had been powered by the Winton 201A, which was one of the first modern compact, lightweight, high-output diesel engines. While it was one of the best engines when it was introduced in 1933, Electro-Motive decided that to successfully pursue the heavy locomotive market it needed a more powerful and rugged diesel. Electro-Motive began design work in 1936, and the 567 engine – a two-cycle "V" design, with welded frames and primary components made from forged castings and steel plate – was ready in 1938. The "567" designation describes individual cylinder displacement in cubic inches. Traditionally, the 567 engine was aspirated using "uniflow" air scavenging forced by a cam-driven Roots blower running at roughly twice engine speed. The engine worked at 800 rpm during maximum output.

Initially, the 567 engine was built in 6-, 8-, and 12-cylinder models. Significantly, Electro-Motive 103 was the first to use the 16-567 configuration. By the time Electro-Motive finally supplanted the 16-567 design more than a quarter-century later, tens of thousands of engines like this were in service across the United States and around the world. Although the basic dimensions and configuration of the 567 engine remained

In 1989, Electro-Motive restored pioneering FTA 103A and repainted former Southern FTB 4103C to match. Union Pacific DDA40X 6936 leads the historic locomotive through California's Feather River Canyon en route to the 1991 Sacramento Railfair. (photo by Brian Solomon)

At Depew, east of Buffalo, New York, in 1946, Lackawanna 603 A, B, and C — an A-B-A set of road service FTs — and a modern 4-8-4 "Pocono" lead an eastward freight. (photo by Joseph Brauner, J.R. Quinn collection)

Southern 4109 leads an A-B-B-A FT set with a tonnage freight at Springfield, Virginia, in March 1949. Southern and its subsidiaries operated 76 FT units. (J.R. Quinn collection)

the same during its long production, Electro-Motive implemented significant design improvements to improve efficiency, reliability, and output. These are designated by letter suffixes, A, B, etc., and are discussed later in greater detail.

The rugged, powerful 567 powerplant was Electro-Motive's key to success with its F-unit and other post-1938 designs. With FT, the 16-567 powered a D-8 main generator providing 600 volts of direct current to four Type D-7b traction motors. These are nose-suspended motors engaging one driving axle using direct gear drive. With this arrangement, the motor is straddled between the truck frame and the driving axle that it powered. One characteristic of the FT design was its mechanical belt-driven auxiliaries. These included an auxiliary generator used to produce between 74 and 78 volts DC, the radiator fans, and traction motor blowers.

The function of a diesel-electric locomotive electrical transmission has been compared to the function of an automobile's mechanical gearbox transmission. The car's engine can't provide torque and horsepower in a speed range wide enough to be connected directly to the rear wheels. So a gear ratio (like third gear, for example) is selected by the driver to match the engine characteristics to the desired output of the car. Likewise for the F-unit, the generator can't provide the current at low speed or the voltage at high speed to be connected directly in one configuration to the traction motors. Transition is used to match the characteristics of the generator to the requirements of the operation of the locomotive. Transitions are accomplished using various series and parallel motor connections. On the FT there were four motor connections — series-parallel, series-parallel-shunt, parallel, and parallel-shunt. The locomotive engineer needed to initiate motor transitions manually using a lever in the cab. A traction-motor transition indicator aided the engineer in timing transitions for optimal efficiency. Transitions were made both as the locomotive

Near the end of their career, this set of Western Pacific FTs was photographed near the Oakland, California, waterfront. Some of WP's FTs survived until 1966-1967, a few years longer than those on most other railroads. (photo by Bob Morris)

Southern Pacific subsidiary St. Louis-Southwestern, known as the "Cotton Belt," bought 20 FT units during World War II. Large illuminated side number boards, as pictured on 921, was a feature of Cotton Belt's FTs in the early years and was also used by Great Northern, Missouri Pacific, Rio Grande, and Santa Fe. (J.R. Quinn collection)

gathered speed (forward transition) and as it slowed (reverse transition).

Blomberg Trucks

Martin Blomberg was hired by Electro-Motive from the Pullman Company in 1935. Among his contributions to diesel design was a successful six-wheel truck, which used an unpowered center axle. This was standard equipment on the Electro-Motive's passenger service E units. He adapted this basic design for use on four-wheel trucks. The FT was the first locomotive model to ride on the new four-wheel Blomberg truck. This proved to be an exceptionally successful design. Most subsequent F units rode on Blombergs, and variations of this truck remained a standard feature on Electro-Motive four-motor road diesels through the 1990s.

Blomberg truck frames were constructed of cast steel alloy. Journal boxes

On June 29, 1945, Baltimore & Ohio FT lead train No. 97 with 92 cars at Engle, West Virginia. These FTs utilize dynamic brake housings — the raised sections on the roof — with tapered sides instead of the more common straight sides. (photo by Bruce Fales, Jay Williams collection)

support truck frames with pairs of coiled springs, while a central swing bolster, used to minimize lateral shocks, is supported on each side with side-hangers holding full-elliptical springs. The side hanger spring assembly situated between the wheels is an easy way to distinguish this type of truck. The axles are centered 9 ft. apart. The FTs used standard 40-in. wheels. A sand pipe is situated on one side of each truck, usually on the leading wheel (based on forward direction), and placed a few inches above rail level. Sand is used to reduce wheel slip and maintain traction in situations with less-than-ideal rail conditions such as rain or if debris or oil has accumulated on the rail head.

The First FT Fleet

An eastward Burlington freight from Galesburg to Chicago passes the tower at Galva, Illinois, on November 25, 1956. Burlington was a natural customer for the FT, having bought Electro-Motive's first diesel streamliner in 1934, the famous Pioneer Zephyr. *The railroad had 32 FT units; all were traded back to Electro-Motive in the early 1960s. (photo by Jim Shaughnessy)*

By 1940, the Atchison, Topeka & Santa Fe had more experience with the operation of diesel locomotives than any other railroad in North America, and quite possibly, the world. Santa Fe was not the first to explore diesel operation — it only began testing diesel operation in 1934. By then some lines, notably Central Railroad of New Jersey, had been operating diesels in switching service for the better part of a decade. Yet in 1940, Santa Fe had a fleet of nearly 60 diesel locomotives, including a significant fleet of Electro-Motive switchers and road passenger locomotives. In November 1940, *Railway Mechanical Engineer* published an authoritative article entitled "The Design, Operation and Maintenance of Diesel-Electric Locomotives." The article described difficulties, challenges, and successes of Santa Fe's brief, but extensive, experience with this new form of motive power. The railroad had assessed the limitations of continuous operation of diesel locomotives, their performance in heavily graded territory, and limitations of the equipment in desert environments. Although Santa Fe had initially bought diesel switchers and passenger locomotives, it was clearly interested in the application of diesels in heavy freight service.

Interest in diesels came from the railroad's top management. In 1939, Fred Gurley came to the Santa Fe from the Burlington, where he had been instrumental in the introduction of the early streamliner *Zephyr*. According to Keith L Bryant, Jr. in his *History of the Atchison, Topeka & Santa Fe Railway*, J.P. Morris from the mechanical department encouraged Gurley and then Santa Fe president, Edward J. Engel, to invest in road diesels. Gurley was a vice president, and during World War II he became Santa Fe's president.

Following extensive testing of Electro-Motive 103, Santa Fe ordered pro-

As built with drawbar connection between A and B units, FTs could be assigned as 2,700-hp locomotives, or as 5,400-hp locomotives. Boston & Maine eastward freight MB-2 passes Concord, Massachusetts, in February 1945. (photo by George C. Corey)

In July 1956, a pair of FTs lead a freight at Apex, New York. This train working over weedy track is going from Norwich and Oswego, New York, to NYO&W's yards at Cadosia, New York. Soon the tracks would be lifted and only weeds would remain. Ten years after this image was made, New York Central used these FTs as trade-in credits for new GP40s. (photo by Jim Shaughnessy)

NYO&W 802 leads a quartet of FTs with an eastward freight approaching Maybrook Yard. Maybrook, New York, was an important freight interchange. NYO&W's FTs were all equipped with dynamic brakes, a feature identified by the raised rectangular roof sections. (photo by Jim Shaughnessy)

NYO&W's FT set 601A and 601B were financed by their participation in the Standard Oil Development Tests. Years after the testing had finished, the famous pair are seen Middletown, New York. (photo by Jim Shaughnessy)

duction FTs. The first of these entered mainline service at Argentine Yard, Kansas City, Kansas, on February 8, 1941. Santa Fe's FTs were different from both prototypes and many production units. The railroad didn't subscribe to Electro-Motive's drawbar connections and wanted its FTA and FTB units equipped with conventional couplers. All its FTs were delivered with couplers, which allowed Santa Fe to arrange them in any combination that it saw fit to use. In the earlier years of operation, Santa Fe routinely assigned FTs in A-B-B sets. Effectively, Santa Fe established operational flexibility of building diesel units to suit tonnage that became the standard for diesel operation in North America.

Another major innovation with Santa Fe's FTs was the pioneering use of dynamic brakes. Reversing traction motor current makes them function as electric generators. This retards forward progress, allowing the motors to effectively work as brakes. Braking current is directed into air-cooled grids to dissipate energy generated by the motors. A similar system, called regenerative braking, had been used on straight-electrics for decades. In that system, the braking current was fed

This rooftop view of NYO&W FTA 803 shows the location of radiators fans, two circular grids behind the horns, and the dynamic brake housing. Four engine exhaust stacks are positioned between the two dynamic brake housings. (photo by Jim Shaughnessy)

Side view of NYO&W FTA 808 provides a good study of cab details. Notice the combined number board and class-lamp fitted to the nose. The rectangular trap below the cab window is used for filling the sand reservoir. (photo by Jim Shaughnessy)

back into the overhead electrical system, which helped balance the cost of operation. On a diesel-electric, the main generator is used to provide field excitation of the traction motors working in dynamic braking mode. The FTs used stainless-steel grids located in the roof, and the fans cooling them were electrically arranged to consume part of the electricity generated by the braking. It is easy to identify FTs with dynamic brakes by spotting the raised boxy section on the roof which houses the grids located toward the middle part of the unit and situated above the row of portholes. The raised section breaks the even contour of the FT's streamlined roofline, making it an aesthetically awkward addition to the locomotive profile.

Santa Fe's early FTs featured a primitive dynamic brake that could only be used at a very limited speed range. Later locomotives came equipped with a more advanced dynamic brake that could be used over a wider range of speeds. The application of dynamic braking proved extremely valuable in train handling and became a great advantage of diesel-electric operations on heavily graded lines. Santa Fe crosses a succession of steep grades on its Chicago-to-California routes and operated some of the steepest mainline grades in the United States, making improved braking a very desirable characteristic.

Dynamic braking simplified train control and greatly reduced dependence on air braking. In some situations the use of train airbrake was cut by 75 percent. This reduced wear on brake shoes, and it largely eliminated the need for heavy trains to stop when descending long gradients to cool overheated brake shoes. Skilled application of dynamic brakes can negate airbrake use in many situations. However, airbrakes are still required when bringing a train to a complete stop. (The nature of the FT's dynamic brake precluded use at very slow speeds.)

Between the end of 1940 and August 1945, Electro-Motive built 320 FT units for Santa Fe, providing an equal number of A and B units, allowing for 80 four-unit sets. This not only gave Santa Fe the world's first fleet of mass-produced road freight diesels, but also the largest roster of FTs. As delivered, they were painted cobalt blue and yellow with the Santa Fe herald on the nose. Locomotives were numbered 100 to 179. The railroad adopted a somewhat confusing practice of numbering the boosters with an "A" suffix, so the "B" unit originally assigned to run with number 100 was identified as 100A. The four-unit set would have been numbered 100, 100A, 101A, 101, since the cab units (A units) would have been at the ends of the locomotive. Adding to further confusion was the fact that later orders included cab units that carried a "C" suffix. One set of FTs was built for dual service. After World War II, Santa Fe temporarily converted additional FTs for passenger service, dressed in the famous red, yellow, black, and silver "Warbonnet" paint livery.

The most significant early application of Santa Fe's FT fleet was on its desert mainlines between Winslow, Arizona, Barstow, California, and mountain crossings between Barstow and San Bernardino, California, over Cajon Pass, and Barstow to Bakersfield, California, over the Tehachapis. During the height of World War II, these routes were saturated

A pair of New York, Ontario & Western FTs lead a freight across the trestle at Ferndale, New York, on March 23, 1957. This view illustrates the asymmetrical arrangement of the FTB. (photo by Jim Shaughnessy)

March 23, 1957, six years after New York, York Ontario & Western FTs 601A and 601B concluded Standard Oil Development Tests and six days before the railroad would cease operations, the pair lead a freight near Middletown, New York. (photo by Jim Shaughnessy)

Boston & Maine's FTAs originally featured a winged banner on the side of the unit such as that seen on No. 4212 at the Boston Engine Terminal in February 1950. (photo by George C. Corey)

with war matériel moving to the Pacific Theater. Santa Fe's desert lines were notoriously difficult. In addition to facing continuously steep grades — including the longest unbroken climb in the United States from the Colorado River crossing at the California-Arizona state line (near Needles, California) to Yampai Summit — trains had to cross many miles of bleached desert that was unbearably hot in the summer and nearly void of usable water for steam locomotives. The little water available was of poor quality for boiler use, facilitating boiler corrosion, foaming, and mineral build-up. Santa Fe had to transport dozens of tank cars of water daily to remote locations to supply locomotive boilers. The servicing facility at Ashfork, Arizona alone received as much as one million gallons a day. Complete dieselization solved these problems, and by 1942 Santa Fe boasted the busiest and longest totally dieselized mainline in the world.

Major maintenance facilities were constructed at Winslow and Barstow, with additional facilities at San Bernardino and Bakersfield. Santa Fe was a rolling laboratory for dieselization and perhaps the best example of what Electro-Motive F-units could accomplish in heavy service.

World War II and the FT

World War II greatly increased railroad traffic nationwide, driving up the demand for motive power — especially Electro-Motive's FTs — but simultaneously limited the availability of crucial resources for non-military use. The War Production Board (WPB) placed strict limitations on what locomotive types could be built and by which builders. Electro-Motive was designated as the sole builder of road-diesels, while traditional steam builders Alco and Baldwin were allowed to build diesel switchers in addition to heavy steam locomotives. During 1942 and 1943 when FT production was suspended, Santa Fe was forced to buy more 2-10-4 Texas types in order to keep pace with its extraordinary motive power needs.

Early on, Boston & Maine made an effort to keep its FTs clean and polished. B&M No. 4217 leads eastward symbol freight MB-2 at South Acton, Massachusetts, in October 1946. (photo by George C. Corey)

The May 1943 *Railroad Mechanical Engineer* featured a General Motors' Electro-Motive Division advertisement that stressed the prowess of its product and GM's role in winning the war. A quartet of shiny new FT diesels were shown hauling a long line of refrigerated box cars on Santa Fe's GFX (Green Fruit Express), and the ad copy read:

"During 1942, with 26 percent fewer locomotives, the Santa Fe moved 122 percent more freight ton-miles and 79 percent more military and civilian passenger-miles than in 1918, during World War 1. All available units of motive power, including supporting facilities, have been and still are being used to the maximum of their capabilities."

In spring 1943, at the height of the war, a General Motors ad featured a pair of resplendent Great Northern FTs, announcing (see sidebar at right):

GM DIESEL FREIGHT LOCOMOTIVES SOLVE VITAL WAR PROBLEMS

Great Northern 2700 hp. Diesel Freight Locomotive No. 5600, operating between Great Falls and Butte, Montana (170 miles), and serving one of the world's greatest copper producing areas, has met every test — solved every problem — with typical-super-performance, as evidenced by the following comparisons with steam locomotives previously used.

Diesel has increased maximum tonnage more than 47 percent on 2.2 percent grade — running time reduced four hours in each direction.

In 62 days of operation Diesel saved 20 round trips — 6,800 train miles; also saved 36 helper trips — 2,103 helper miles. Helper service was entirely eliminated on a 54-mile district and reduced 50 percent on a 10-mile district.

GM Diesel efficiency is not affected by altitude or extreme temperatures.

GM Diesel had eliminated smoke and gas hazards in tunnels and fire hazard along right of way.

Why did the Great Northern adopt GM Diesels? The abnormal demand for copper, one of the most vital war materials, required either rebuilding the line to accommodate heavier steam power or adopting GM Diesels, which had already demonstrated their ability to outperform steam without expensive changes in operating facilities. GM Diesel locomotives positively reduce wear and tear on track and bridge structures. To even approach this Diesel performance with steam power would have required an expenditure of $350,000 to $500,000 for reducing grades, reinforcing bridges, enlarging tunnels, and installing a new turntable — an amount considerably more than the cost of the Diesel.

Transportation is Vital for Victory

On January 12, 1957, B&M FT No. 4212 leads an eastward freight past Hoosic Falls, New York, on grade-separated track. The westward mainline crosses over the top of the locomotive. (photo by Jim Shaughnessy)

Like most FT sets, Boston & Maine's feature a semi-permanent drawbar connection between the A and B units. This pair leads a freight at Northampton, Massachusetts, on August 11, 1946. (photo by Robert A. Buck)

The FT's signature carbody feature was the row of four tightly spaced porthole windows. (photo by Jim Shaughnessy)

Boston & Maine's first FT at Mechanicville, New York, in March 1957. This locomotive was built in 1943 and traded in on GP9s in summer 1957. The multiple-bulb headlight was an Electro-Motive option, exercised by B&M on many of its early carbody units. (photo by Jim Shaughnessy)

An engineer climbs on board Boston & Main No. 4222 at Mechnicville, New York, in 1957. The FT's Blomberg trucks used an older style of journal than what was used on later F units. (photo by Jim Shaughnessy)

In this ad, General Motors was not just trying to impress railroad management by boasting about the FT's achievements, but they were also addressing the curtailment of FT production because of need for copper in its electrical system. GM was feeling the pinch of the WPB at the very moment when its product was in highest demand.

After World War II, Electro-Motive was free to expand production and implement improvements. On Santa Fe, as the tide of war traffic subsided, the railroad greatly expanded FT's working territory. In addition to the passenger assignments already discussed, FTs regularly worked eastward to Argentine Yard in Kansas City.

FT Tames the West

The FT influence covered most regions of the United States from New England to California and from the Deep South to the Pacific Northwest. Yet the largest numbers of units were sold in the far West. Western railroads operated three of the four largest FT rosters, and many western railroads sampled the FTs. As noted, Santa Fe's 320-unit fleet alone accounted for 30 percent of the total pro-

Boston & Maine No. 4200 looks faded and worn as it leads a southward Connecticut River Line freight at East Northfield, Massachusetts, on May 7, 1955. (photo by Jim Shaughnessy)

Horns blaring and tonnage in tow, Boston & Maine No. 4205 races east through Buskirk, New York, in 1955. The first porthole window on an FTA was hinged for access. (photo by Jim Shaughnessy)

ELECTRO-MOTIVE FT

Boston & Maine FTA No. 4221 was involved in a wreck at Biddeford, Maine, that damaged its nose. The locomotive was rebuilt with an F3-/F7-era nose section with the later style of large, boxy number boards. (photo by Jim Shaughnessy)

This down-on view at East Northfield, Massachusetts, gives a good view of radiator exhaust vents, dynamic brake housing, and horns on a B&M FTA. Leading this set is an F2A bought in 1946 to augment the FTs. (photo by Jim Shaughnessy)

duction run. Great Northern had the second largest, with 96 units, followed by Southern Railway and its subsidiaries with 76 units. Milwaukee Road had 52 FTs. By the end of World War II, the sounds of 16-567s could be heard roaring over Tennessee Pass and up the Front Range on Denver & Rio Grande Western, crossing Washington State's Stampede Pass on Northern Pacific, racing across Kansas wheat fields on Rock Island Lines, stirring up the dust in Texas on Missouri Pacific, while rolling "Everywhere West" on Burlington.

Unlike steam power, which for decades had been dressed in basic black, FTs were painted in a great variety of liveries, most dreamed up by Electro-Motive designers. These bold new locomotives in bright and shining paint would have been a thrill to see during the uncertain days of World War II. Electro-Motive placed big color advertisements in the railroad trade press boasting of the FT and its capabilities. One ad depicted a bulldog grimacing as it tugged on a heavy rope which was over-laid on a four-unit Santa Fe FT set. The headline read in bright letters "The Bulldog of the Rails," which played on the aesthetic similarity between the FT nose and the Bulldog's profile.

"Like the proverbial bulldog which pulls and pulls and never quits — the EMC 5,400-hp Diesel, the world's most powerful freight locomotive, can outpull and outperform any steam locomotive thus far built. This 'Bulldog of the Rails' has greatest tonnage moving capacity, can operate continuously over long runs, and with its dynamic braking feature can handle trains down heavy grades faster, safer and with minimum brake applications — and at a greatly reduced cost of operation.

"On a recent test trip between Argentine, Kansas, and Los Angeles, with no attempt for speed or tonnage records, Santa Fe 5,400-hp Diesel freight locomotive No. 100 with a maximum of 68 cars (3,150 tons) made the 1761.8 miles in 54 hrs. 35-1/2 mins. running time — an average speed of 32.3 m.p.h."

Since that time, this locomotive profile has been known colloquially as Electro-Motive's "Bulldog" cab. This style was applied to all the F units as well as postwar E units — models E7 to E9 — as well as a variety of models (assembled) overseas, such as Australia's famous A7.

In a June 1944 GM advertisement, Milwaukee Road was used as an example of the FT's performance. Milwaukee operated the longest mainline electrification in the United States, totaling more than 650 miles across the Rocky, Bitter Root, and Cascade Mountains. It had made a pioneering investment in overhead electrification during the World War I period when that technology seemed to be the most efficient way to move tonnage. Milwaukee's electrification had two long segments separated by a 226-mile gap between Avery, Idaho, at the western end of its Rocky Mountain electrification, and Othello, Washington, at the eastern end of its Cascade electrification. This had caused an operational bottleneck, as it was possible to operate much longer and heavier freight trains with electric power than possible with steam. As a result, heavy trains needed to be reduced (shortened) for operation between Avery and Othello. It was on this section that Milwaukee assigned some of its first FTs.

A February 1947 article in *Diesel Railway Traction* highlighted the

Locomotive No. 5414, seen at Alamosa, Colorado, is a typical example of how Rio Grande modified its FTs in later years. Notice the boxy homemade number boards and the unusual class lamp arrangement. These were the original class lamps cut free from the old number boards and welded in this new position. (photo by Jim Shaughnessy)

specifics of Milwaukee's unusual electric-to-diesel-back-to-electric traffic hand-off. Between 1932 — at a traffic low during the Great Depression, Milwaukee's Pacific Extension (its lines to Seattle and Tacoma) — and 1945 the railroad's freight traffic, measured in gross ton miles, had more than quadrupled. So while this had been a difficult operation situation prior to the War, it became a traffic manager's nightmare as traffic levels ballooned. Milwaukee had assigned some of its largest steam power to haul freight over the non-electrified gap. These were its Class N3 2-6-6-2s, massive oil-burning articulated locomotives with an 82,720-lb tractive effort rating. Yet, these behemoths were no match for a four-unit set of new GM diesels, which delivered 230,800 lbs. tractive effort, allowing them to take over trains directly from Milwaukee's gargantuan General Electric boxcab electrics without need to reduce tonnage.

FT for Freight

Baltimore & Ohio was another early purchaser of Electro-Motive products. It was the first E-unit customer, and was a logical early customer for the FT. B&O was the first railroad in the east to buy FTs. Its first four-unit set carried road-number 1 and was delivered to the railroad at Chicago on August 31, 1942. According to the October 1942 *Railway Mechanical Engineer*, on that day it was dispatched from Barr Yard with 76 loaded oil-tank cars for an 805-mile nonstop run to Baltimore, Maryland. The symbolic and strategic significance of the movement of oil by rail was reported in the railroad press. Obviously, oil was the lifeblood of the diesel-electric, but perhaps more importantly, during the War, a German naval blockade of the Atlantic Coast had disrupted the fuel supply, which was normally transported by ship. To counter this threat, oil was shipped inland by rail, and solid trains of oil became a common sight on American rails.

In his *History of the Baltimore & Ohio Railroad*, John Stover puts perspective on the magnitude of B&O's oil shipments: In 1941, prior to the blockade, total rail shipments of oil to the East Coast amounted to 11,250 barrels a day. By 1943, this had jumped to more than one million barrels daily. In that year, B&O was moving an average of 850 cars of oil a day. In 1944, oil accounted for more than 3 percent of B&O freight tonnage. A November 1942 Electro-Motive Division advertisement boldly featured B&O No. 1, explaining that this new 5,400-hp diesel had hauled a "solid train of 81 tank cars of oil — 715,000 gallons — 5,300 gross tons — from Chicago to Twin Oaks (Philadelphia), Pa." This run covered 911 miles and notably required only two stops for refueling.

Between 1942 and 1944, Electro-Motive delivered 24 FT units. Six of these were four-unit sets to B&O. Stover notes that B&O's FT sets cost about $500,000 each. B&O's moderately sized FT fleet is a classic example of how the War dramatically boosted traffic, creating a demand for diesel power, while the limitations of the WPB prevented the railroad from getting as many FTs as it wanted. Instead of more FTs, the WPB encouraged B&O to

Great Northern was among the earliest railroads to buy FTs — the first arrived on the property in May 1941. In this July 20, 1955, photo, FTs are seen eastbound at Cass Lake, Minnesota. (photo by Jim Shaughnessy)

Boston & Maine's pioneer FT meets the new guard at Mechanicville in 1957. The FT's belt-driven appliances made upgrading a costly prospect. Electro-Motive offered the option to recycle FT components on new GP9s, giving customers a considerable savings. (photo by Jim Shaughnessy)

order modern steam instead. In 1944 and 1945, Baldwin built B&O's famous Class EM-1s, simple-articulated locomotives with the 2-8-8-4 wheel arrangement. These 30 locomotives were its last new steam power. After WPB restrictions were eased, B&O, like most American railroads, began ordering new diesels at an accelerated pace to replace worn-out steam power.

Since it operated numerous mountain grades on its eastern lines, B&O's FTs were logically equipped with dynamic brakes. Rosters indicate that B&O's FT A-units weighed 229,500 lbs. and delivered 57,373 lbs. starting tractive effort, while the FT B-units were nominally lighter, weighing 228,000 lbs., delivering 57,000 lbs. starting tractive effort. As built, they were geared for 70 mph operation. During B&O's FT production, Electro-Motive introduced the improved 16-567A engine, so some of its locomotives had the older engine model and others the newer one. Diesel Era's *The Revolutionary Diesel* indicates that this resulted in different classes, as FTA/FTBs with the older 16-567 engines classed DF-1/DF-1x and the 16-567A DF-2/DF-2x. Baltimore & Ohio was not alone in having FTs that straddled the change in engine models. Santa Fe also had FTs with both engines. This relatively minor alteration is noteworthy

The bankrupt New York, Ontario & Western bought FTs to cut costs. Streamlined Electro-Motive diesels couldn't help the ailing carrier. A quartet of FTs lead a westward New York, Ontario & Western freight near Cooks Falls, New York, on July 8, 1956. (photo by Jim Shaughnessy)

because the FT model underwent very few changes during its six-year production, largely because of WPB limitations. *The Revolutionary Diesel* also notes that B&O later replaced the semi-permanent drawbar connections between FTAs and FTBs with more conventional couplers. In 1948, the FTs were renumbered from 1 to 11 as built, to the 101-series, and a decade later they were renumbered again to the 4400-series for A units and 5400-series for B units.

In total, 23 railroads bought the new FTs. Although less numerous in the East than in the West, many eastern lines, including most of the "Anthracite Roads," sampled FTs. The Delaware, Lackawanna & Western operated from Hoboken, New Jersey (across the Hudson from New York City) to Buffalo, New York, by way of Scranton, Pennsylvania, and Binghamton, New York. In 1944-1945, it bought two fleets of FTs, totaling 24 units. One fleet, numbered 601-604 (A, B, C) intended for three-unit A-B-A sets in road freight service, used the common 62:15 gearing, while the others consisted of A-B sets numbered 651-654, which employed 65:12 gearing for slow-speed, high-tractive effort work as helpers based out of Scranton. Lackawanna's chief competitors, Erie Railroad and Lehigh Valley, also purchased FTs.

Erie Railroad initially assigned its FTs, numbered 700-705 (A, B, C, and D) in four-unit sets to work road freights on its Kent and Mahoning Divisions between Meadville, Pennsylvania, and Marion, Ohio. In *Men of Erie*, Edward Hungerford explains that Erie's route across northern Ohio between its major yards at Meadville and Marion was a series of 1-percent grades in a "saw tooth" profile. As a result, steam-hauled eastward freights coming from Chicago would have to reduce from 5,000 tons to about 3,000 tons at Marion. FTs solved this problem with their greater starting and pulling power and dynamic braking. Hungerford cites an expenditure of $4 million (including $1 million for the construction of a diesel shop at Marion)

Although New York Central was the largest railroad to purchase FTs, it had just eight units, one of the smallest fleets. On October 11, 1952, an FT pair lead an eastward freight at west of Athol Springs, New York. (photo by S.K. Bolton, courtesy of George C. Corey)

as the cost to Erie for this dieselization. At the end of World War II, Erie's Marion yard was accommodating 52 heavy road freights daily. By the early 1950s, Erie was the first of the four major east-west trunk lines (the other trunks were New York Central, Pennsylvania, and B&O) to completely dieselize its road freight operations.

Anthracite coal hauler Reading ordered a fleet of 20 FT units that initially worked in ten A-B pairs, numbered 250 to 259. These pairs weighed 457,700 lbs. and delivered 114,425 lbs. starting tractive effort.

FTs for the Old & Weary

One of the most fascinating applications of new diesels was on the New York, Ontario & Western, which operated a beleaguered network of lines between Weehawken, New Jersey, Scranton, Pennsylvania, and the Lake Ontario port of Oswego, New York. This line had long been among the weakest railroads in the Northeast. Until the late 1930s, it had survived primarily on revenues from lucrative coal traffic originating in the vicinity of Scranton. On the eve of World War II, NYO&W's coal traffic evaporated and the railroad descended into bankruptcy. During the War, NYO&W's trustees decided to try to keep the line in business by transforming it into a modern bridge line, and they authorized the purchase of new diesels and installation of Centralized Traffic Control on key routes. NYO&W bought FTs not just to improve its operations, but also to lower its costs to stay in business. The railroad was on the brink of oblivion and hoped diesels might save it. The railroad was not among those that had hosted demonstrator No. 103. It based its decision to buy FTs on inspection of other railroads' FTs, Electro-Motive literature, and interviews with railroaders familiar with FT performance. In deciding to purchase FTs, NYO&W was not just sampling Electro-Motive's products, but planning to replace its steam fleet. Making the decision to dieselize operations at this time was a bold and expensive move and was not taken lightly by railroad officials. NYO&W was dieselizing to survive. Diesels, in conjunction with other technology such as the installation of CTC,

Anthracite coal-hauler Reading normally operated its FTs in A-B pairs, using them on the head end of heavy freights and as helpers. All of them were traded back to Electro-Motive for new GP30s in 1962. (photo by Richard Jay Solomon)

were part of its strategy to transform itself into a modern route that would forward traffic between other railroads.

New York, Ontario & Western was the last railroad to decide upon the FT, although other lines that already owned them re-ordered before Electro-Motive ceased production in 1945. NYO&W's FTs arrived in spring 1945, having worked their way east over the Erie Railroad. The first revenue operation occurred in mid May. NYO&W operated a fleet of 18 FTs, arranged in nine A-B pairs. In his book, *New York Ontario & Western in the Diesel Age*, Robert E. Mohowski indicates these cost an estimated $2.2 million. The FTs were part of a strategy that the railroad's trustees believed would save approximately $1 million per year over steam locomotive operation. Nonetheless, raising the money for acquisition of the FTs was difficult for the financially strapped line. Mohowski relates that NYO&W paid for its diesel purchases in part though the sales of passenger equipment, old steam locomotives, and scrapped rails from the single tracking of its once double-track line between Cornwall, New York, and Mayfield, Pennsylvania. In addition to the cost of diesels, the railroad also needed to build specialized fueling and servicing facilities.

One set of FTs, numbered 601, was paid for by a loan from the Standard Oil Development Company. According to Mohowski, who has written extensively about these locomotives, the SODC worked with NYO&W to perform lubrication and fuel-oil tests to perfect its products for railroad diesel applications. As a result, 601's service was closely monitored and primary engine components such as cylinder and cylinder liners were removed, inspected, and replaced between tests. Testing was conducted over a six-year period ending in 1951.

NYO&W's FTs were painted in a distinctive livery designed by Electro-Motive. Locomotive bodies were bright gray with a broad yellow stripe that wrapped around the sides, running just below the porthole windows, and swinging up across the nose. On the nose, just below the broad yellow stripe, was a swatch of orange that arced down and ran along the top of the anticlimber. Thin yellow stripes accented the lines of locomotive sides just below the air intakes. The railroad's distinctive logo, a stylized "W" encircled by an "O," was boldly displayed in red paint directly below the headlight within the yellow stripe. This livery had decidedly autumnal qualities, which complemented the scenery the railroad traversed, but in retrospect also seemed to reflect the condition of the railroad in its declining years.

New York, Ontario & Western primarily assigned FTs to road freights working between New York's Maybrook Yard — an important interchange point on its eastern end — and Mayfield Yard in the Scranton area. FTs also worked freights to Weehawken, New Jersey, using trackage rights over New York Central's West Shore line south (New York Central direction east) of Cornwall.

Despite its efforts to modernize, NYO&W fought a losing battle. Diesels and CTC signaling could not save it, and in its last years the railroad was characterized by these streamlined diesels wading through weed-covered tracks. In 1957, the railroad ceased operations and most of its lines were abandoned. Its diesels were sold, and two sets of FTs, Nos. 806 and 807, found new life on Baltimore & Ohio. The remaining units were bought by a locomotive dealer and were eventually acquired by New York Central to use as trades back to Electro-Motive.

FTs in the Berkshires

Like Santa Fe and other railroads interested in the FT, New England-based Boston & Maine was already well acquainted with Electro-Motive locomotives by the time No. 103 debuted. In 1935, B&M had bought an Electro-Motive-powered Budd streamliner, operated as the *Flying Yankee*, that was a near copy of Burlington's *Zephyr*, and beginning in 1936 it had sampled various switcher types.

In September 1940, No. 103 graced the rails of the Boston & Maine working freights on its Fitchburg Mainline. According to the October 1940 *Railway Mechanical Engineer*, on September 5th No. 103 made its first run on B&M, and immediately set a time record hauling an 83-car, 4,500-ton freight between Mechanicville, New York, and B&M's Somerville Yard near Boston. In addition to their great pulling power and other efficiencies, diesel operations appealed to

B&M because of their ability to operate through the 4.75-mile-long Hoosac Tunnel without assistance from electrics. B&M had electrified its Hoosac Tunnel operations in 1911 to increase capacity and reduce problems caused by smoke and gases accumulating in the tunnel.

In 1943 and 1944, Boston & Maine acquired a fleet of 48 FTs. These were dressed in a handsome maroon paint with gold striping in a livery designed by Electro-Motive, which was very similar to that also used on Rio Grande's FTs. The A units had Boston and Maine spelled in bold type along the sides below the four-porthole windows. They were numbered in the 4200 series. One unusual characteristic of B&M's early carbody units was B&M's choice of a multi-lamp headlight that featured a ring of six lamps around a central lamp.

According to *The Revolutionary Diesel*, B&M's FTs were rated at 7,400 tons westbound between Boston and its yards at East Fitchburg. This was roughly double that of its most modern freight steam. Initially B&M assigned its FT in A-B-B-A sets. However, this often provided more power than the railroad actually required to move tonnage, and FTs were sometimes split up into two-unit A-B sets. After the war, B&M bought F2s (discussed in the next chapter) in order to run A-B-A sets, with a single F2 on one end of an FTA-FTB combination.

Boston & Maine's postwar traffic dropped off precipitously. In 1946, it completely dieselized operations through the Hoosac Tunnel, thus ending the need for electrification. To make this possible, a modern ventilation system was installed in the tunnel to keep it free from diesel exhaust. In the mid-1950s, B&M was one of the first lines to retire its FT fleet, as it traded them back to Electro-Motive for a new fleet of 50 GP9 road switchers constructed during 1957.

Electro-Motive Styling and Painting

Electro-Motive moved its locomotive styling and section to La Grange in about 1944, as post-World War II production was almost ready to swing into high gear. It was a busy time for the diesel builder. Design patents show 16 different paint schemes for freight units through 1946. Most were credited to John Markestein, who came to the U.S. from Holland in the 1940s. Markestein, one of the originators of the Dutch airline KLM, was an engineer. He supervised the styling section, since styling was a part of the engineering department.

Markestein (or his staff) prepared the schemes for Rio Grande, Boston & Maine, Cotton Belt, New York Ontario & Western, Chicago & North Western, Reading, Minneapolis & St. Louis, Erie, Lackawanna, Maine Central, Northern Pacific, and Western Pacific. Earlier, from Detroit, came styles for Southern, Santa Fe, Great Northern, and Chicago Burlington & Quincy.

Markestein's influence was felt for years as the people he hired continued to make styling and painting decisions. They included Harry U. Bockewicz, a commercial artist; Ben Dedek, a design and scenery artist; Ed Moreau, from engineering; Barbara Luse, a draftsman; and Rex Prunty, artist and draftsman.

In 1949, Lee Buchholz was the fifth person to join the staff. Markestein was "a real interesting guy. He hired me and took me under his wing. He taught me a lot, gave me a lot of breaks. I owe my career to him," said Buchholz, who became styling and painting supervisor in 1961.

Markestein "was very much into raising tulips. Every year he would plant different designs. Other people requested his bulbs, so I guess he came up with some new plants."

"When he was supervising the styling section, Markestein did little if any of the actual work, but his initials are on almost every sketch and finished piece. His influence undoubtedly contributed to graphic discipline and the uniform quality of the end product," explained Jim Boyd in his two-part article in *Railfan & Railroad* in November and January 1985.

About 1953, Markestein left La Grange with others in the engineering staff to go to the locomotive plant at London, Ontario. He returned to La Grange after three years, retiring from EMD there. Markestein and Buchholz stayed friends; Buchholz attended his wake at a La Grange funeral home.

"I was there at the peak of production: 11 units a day, eight in La Grange and three switchers in Cleveland," Buchholz continued.

"I did a lot of locomotives in the 38 years I was there. In the last years, 1967 to retirement, I was head of styling and export shipping, and inspected projects in the field — busy enough I didn't need any more jobs.

"Once the group was established, almost all of the artwork and styling and painting responsibilities were handled from La Grange, without Detroit involvement," said Buchholz, who retired in 1986. "We were proud of our accomplishments."

by John Gruber

CHAPTER TWO

POSTWAR FS

Bangor & Aroostook was one of the last to operate F3s in regular revenue freight service. Several units worked into the early 1980s. F3A No. 40 leads a consist of Fs and BL2s at Brownville, Maine, in June 1980. (photo by Don Marson)

World War II had a profound effect on locomotive development, refinement, and application in North America. While a few railroads had dabbled with diesels prior to the war, steam was still king. Electro-Motive's FT had made its debut on the eve of the War and had conclusively demonstrated the merit of high horsepower road diesels in freight service. During the war, production FTs rolled off millions of miles in revenue service, giving railroads hands-on experience with diesels, and further proving their merit while providing Electro-Motive engineers with a wealth of practical knowledge. Working with diesel locomotives in a real-life application gave engineers much greater insight into the locomotive's performance and reliability capabilities, yielding knowledge required to greatly improve their designs. As explained in the April 1946 issue of *Diesel Railway Traction* describing locomotive maintenance on the Rio Grande, "certain [locomotive] faults... can be detected only when the engines are working under load, and perhaps only after some continuous period of full load operation." Heavy freight service in the mountains is the best laboratory for studying locomotive reliability.

All aspects of locomotive performance reliability were studied, as were locomotive components such as the primemovers, main generators, and traction motors. Identifying why these components failed as a result of service allowed Electro-Motive engineers to refine designs, or in some cases come up with new designs, that would last longer in the rigors of railroad service.

The war had other crucial implications for locomotive production. It stifled railroads acquiring new diesels while

In the early 1990s, Bangor & Aroostook restored its last F3A No. 42 to its original number and paint livery. The locomotive was seen at Oakfield, Maine, on May 29, 1994, with an excursion train. (photo by Brian Jennison)

The F2 and the earliest F3s featured three porthole windows on the sides, with the center porthole hinged. Boston & Maine F2A No. 4256 was photographed at night in Mechanicville, New York, in November 1957. 4256's engine room lights are illuminated, giving a hint of the locomotive's inner machinery. (photo by Jim Shaughnessy)

saturating lines with heavy traffic, accelerating wear and tear on the fleets of already tired steam locomotives. Since World War II came on the heels of the Great Depression, many railroads were caught short when it came to motive power. Relatively few new steam locomotives had been built during the 1930s, resulting in a high average age of motive power. Following the attack on Pearl Harbor, the nation mobilized for the war effort, which greatly increased railroad traffic and demands on locomotives. While railroads were now in desperate need for new power, the War Production Board had placed strict limitations on locomotive production, controlling the numbers and types of locomotives available for purchase. Electro-Motive was strictly limited to the production of its FTs, while the other two major diesel builders were basically

This view of NYO&W F3A No. 822 gives a comparison of the differences between the F3 (phase II) roofline and that of the FT. Notice the twin dynamic brake exhaust vents behind the horns, followed by a row of non-shrouded radiator fans. (photo by Jim Shaughnessy)

New York, Ontario & Western F3A 502 has some work done to it at the railroad's company shops in Middletown, New York. After NYO&W's demise in 1957, this locomotive worked in California for Sacramento Northern. (photo by Jim Shaughnessy)

limited to the production of switchers. Furthermore, the WPB discouraged implementation of new designs that might complicate parts supply. In short, innovation was discouraged. Also, since diesel-electric locomotives used crucial materials needed for war machinery such as copper wire and diesel engines, the overall production of diesel electrics was curtailed. Railroads were encouraged to order steam locomotives, even when they would have preferred diesels.

While the implementation of new designs was generally discouraged, research and development was not. The war gave Electro-Motive an unusual opportunity to refine its technology in an environment effectively free from competition, while simultaneously allowing its customers to experience and comment on its product without the ability to demand it.

C. R. Osborn, General Motors vice president and general manager of its Electro-Motive Division, explained in an interview in the October 26, 1946, *Railway Age*,

"[Locomotive] development was dammed up between 1941 and 1945 by the war, when we were busy with large

POSTWAR Fs   31

Railroads had an option of a model identifier plate on most postwar F units. An eastward piggyback train at Galesburg, Illinois, prepares for its journey to Chicago on November 24, 1956. (photo by Jim Shaughnessy)

New York Central F3As Nos. 3500-3503 and F3Bs Nos. 3600-3601 used 56:21 gearing for dual freight and passenger services. In February 1949, an A-B-A set of F3s leads the New England States at Framingham, Massachusetts. (photo by George C. Corey)

Maine Central F3A leads a mix of first-generation Electro-Motive diesels at Bangor, Maine, on April 10, 1955. This unit was a typical "phase II" F3A. (photo by Jim Shaughnessy)

U.S. Navy contracts for our Diesel engines and other equipment, and were restricted to the production of our freight locomotive only.

"By the time we resumed full production we had 238 passenger units in service for years. They had operated a total of more than 200,000,000 miles of regularly scheduled high-speed service... Added to this, we had the experience gained from the operation of more than 1,100 of our 1,350-hp freight locomotive units [model FT]... These freight locomotives have handled more than 160 billion ton-miles of freight."

Osborn elaborated, "Our engineers have had a wealth of opportunity to determine the characteristics of a locomotive which would most nearly meet the requirements for an all-purpose, mainline locomotive capable

Bangor & Aroostook bought eight F3As. These regularly worked with the railroad's other first-generation locomotives. No. 43 switches at the Canadian Pacific yard at Brownville Junction, Maine, in April 1955. Its four F3Bs were sold to the Pennsylvania Railroad in 1952. (photo by Jim Shaughnessy)

On December 14, 1980, Bangor & Aroostook 42 leads the company's annual Turkey Train at Brownville, Maine. This locomotive was restored a decade later to its original livery. (photo by Don Marson)

Bangor & Aroostook F3A No. 41 gets a steam cleaning at Northern Maine Junction in February 1970. Notice the decorative roof overhang at the rear of the locomotive. (photo by Brian Jennison)

Bangor & Aroostook's F3As all used the "phase II" carbody with wire mesh covering air intakes at the top and center sections of the carbody sides. F3As 40 and 45 were photographed at Houlton, Maine, on September 10, 1975. (photo by Brian Jennison)

Form versus function — Electro-Motive models F3A, GP7, and BL2 are seen hauling train No. 44 south of Oakfield, Maine. All three use the same diesel engines and electrical equipment; locomotive configuration is the only real difference between them. (photo by Robert A. Buck)

As good as the FT had been in comparison with steam power, it was far from perfect. Its various idiosyncrasies and failings had been clearly revealed during the pressures of wartime service. Today the old FT appears clunky, underpowered, and antiquated in comparison with later Electro-Motive Fs.

Another postwar consideration for Electro-Motive was the changing face of its competition. In 1939, its FT was offered as an alternative to steam power. After the war, Alco and Baldwin both introduced their own road diesels, and diesel-engine manufacturer Fairbanks-Morse also entered the road locomotive market. While these builders largely followed patterns for diesel-electric locomotive construction established by Electro-Motive, each also looked to improve upon what Electro-Motive had accomplished. So diesels were competing against diesels in a new, rapidly expanding, and very competitive locomotive market. In addition, the locomotive field was still open, builders were still promoting the most advanced reciprocating steam locomotives ever conceived, and various turbine designs were under development.

Yet, at the end of the war Electro-Motive's road diesels clearly had the edge in development, marketing, and existing customer base. However, it needed to continue to improve its products in order to stay abreast of new competition.

Canadian National's U.S. subsidiary Grand Trunk Western had 22 F3As sequentially numbered after CN's own. Two F3As bracket an F7A on a GTW freight in the summer of 1964. (photo by Richard Jay Solomon)

of excelling previous performance of any type of locomotive throughout its wide range of speeds."

As the conflict subsided, WPB restrictions were eased, and Electro-Motive was able to implement design changes while taking orders from railroads hungry for new motive power. Having enjoyed four years of development with minimal market pressures, Electro-Motive now stood poised ready to build new locomotives. Electro-Motive incorporated many refinements in postwar models based on its extensive wartime experiences. Its F-unit line was a prime example.

Standard Products

Electro-Motive offered locomotive models similar to the way parent General Motors offered car models. This was an intrinsic part of Electro-Motive's market

In autumn 1963, former Lackawanna F3As operate a passenger excursion for successor company Erie-Lackawanna in northern New Jersey. (photo by Richard Jay Solomon)

orders. Within the standard models, minor variations were offered to best suit an individual railroad's needs. On F units, variations and options included external equipment such as headlights, horns, and pilot styles, and internal equipment such as dynamic brakes and steam boilers. Primary components, horsepower output, and locomotive configuration remained the same. When Electro-Motive typically introduced new models, it was in conjunction with upgrades to its whole locomotive line. With the F units, the postwar models supplanted the FT and each new model effectively supplanted the last. New models typically incorporated technological changes aimed at improving performance, reliability, and ease of maintenance.

An important element of Electro-Motive's design strategy was maintaining compatibility between model types so that older locomotives could operate with newer ones, and so that most of the parts supply remained constant. Wherever possible, Electro-Motive made primary components interchangeable — making it easier for railroads to upgrade older locomotives to the latest specifications.

strategy. So instead of adhering to the traditional steam-era approach of tailoring locomotive design to individual railroads, it offered just a few standard-production models. Electro-Motive engineers designed what they believed was the most effective locomotive with the intention of mass production to fulfill its customer

Erie-Lackawanna 6212 is an older F3B, built as Lackawanna 621B in January 1948. It was photographed at Buffalo, New York, on July 18, 1971. (photo by Doug Eisele)

A contrast in power and style, Erie-Lackawanna F3A 6621 leads two SD45s on a 93-car westward freight at Hornell, New York, on May 31, 1971. The former Lackawanna F3A has just 1,500 hp, compared to 3,600 hp for each SD45. (photo by Doug Eisele)

The differences that delineated one F-unit model from another were improvements to primary components, including engines, generators, traction motors, and airflow systems. Improvements facilitated better performance through higher horsepower output, increased traction, improved short-time motor ratings, and enhanced reliability or eased maintenance.

This has made identifying and classifying F-unit models more challenging, since external changes do not necessarily indicate differences between models. Furthermore, some railroads upgraded older units to more modern standards. What was built as an F3A may have been upgraded to F7A or F9A standards. Sometimes this resulted in noticeable external modifications, other times not. So simply

New York Central 1620, built in July 1947, is considered a "phase II" carbody. Notice the wire mesh between the two widely spaced side porthole windows. This particular unit has a flush "passenger" pilot, which was available as an option on later Fs. (photo by S.K. Bolton, courtesy of George C. Corey)

Rio Grande F3A 5521 pauses with the California Zephyr at Grand Junction, Colorado, on Aug 4, 1963. This locomotive exhibits F3A "phase III" side panel traits. Note the metal louvers between the porthole windows. (photo by Jim Shaughnessy)

Erie Railroad 708A, B, C, and D, an as-built A-B-B-A F3 set, leads an eastward freight across the Chemung River at Elmira, New York, on September 7, 1956. (photo by Jim Shaughnessy)

Working hard upgrade with its 16-567Bs engine roaring, Southern Pacific F7A 6208 approaches the siding at Mott Railroad, east of Dunsmuir, California. Note the stainless-steel radiator intake grills and four sets of air-intake louvers between the portholes. This is a "phase I" F7A. (photo by Bob Morris)

identifying a locomotive based on its as-built carbody may be misleading.

In some situations, Electro-Motive incorporated performance and reliability improvements prior to introducing a new model designation. Model designations are in part an advertising ploy, and a new model can simply function as recognition of improvement to a basic design. With each *new* model Electro-Motive would be sure to stress all of the latest improvements featured, even when these were options on previous models. One particularly difficult area for observers is the transition between the F3 and F7 models. There were four discernable F3 carbody phases, and the later F3s appear nearly identical to F7s. However, the

Erie Railroad No. 711A is a classic "phase I" F7A. This carbody style was also used for the last F3s, considered F3 "phase IV." The engineer of this train is neither aware nor concerned about carbody phases; to him the beefier traction motors probably matter most. (photo by Jim Shaughnessy)

defining features between the F3 and F7 were fundamental technological advancements not evident in shape of the carbody. Details of carbody changes are discussed later in this chapter. Understanding the technological advances as well as nominal external alterations helps put the evolution of Electro-Motive's F-unit line in a clearer perspective.

Gulf, Mobile & Ohio F3A 884A backs a Chicago-to-Joliet, Illinois, local train into Union Station. This is considered a "phase II" F3A; note the shrouded radiator exhaust fans. It was later rebuilt as MBTA FP10 1151, and today it works on the Adirondack Scenic Railroad as No. 1502. (photo by George Kowanski)

Models F2 and F3

Electro-Motive looked forward to brisk locomotive sales as World War II concluded and the rescinding of WPB restrictions allowed it to finally implement technological advancement. In preparation for postwar sales, in August 1945 Electro-Motive debuted a new four-unit demonstrator initially designated Model F2. Like its predecessor, FT 103, this locomotive was dispatched on a nationwide tour of the United States and ultimately clocked more than 110,000 miles.

The new demonstrator showed off Electro-Motive's much improved diesel design, aimed at convincing skeptical managers that investment in diesels was worthwhile, and giving railroads a preview of the new technology. While the superiority of diesel-electrics over steam seems obvious today, it is important to remember that new diesels came at a much higher cost than steam. A new four-unit FT set was roughly three times that of comparable steam locomotives.

Railroads needed clear demonstration of diesel economy before they were willing to invest in the new technology. For example, Southern Pacific had refrained from buying freight road diesels prior to World War II and remained loyal to steam during the war years. It had ordered new Baldwin "cab-ahead" articulateds as late as 1944 (although its Cotton Belt subsidiary sampled five four-unit FT sets). John Bonds Garmany explains in his book, *Southern Pacific Dieselization*, that in 1945 Electro-Motive offered SP the use of the four-unit demonstrator. This was assigned to heavy service on the Sunset Route between Yuma, Arizona and Lordsburg, New Mexico. In a single month of testing it chalked up 10,000 miles of service. Southern Pacific was impressed. Garmany quotes James W. Corbett, SP's vice president of operations:

"After our experience with the experimental 6,000-horsepower Diesel in freight service and our studies confirmed its operation advantages, Southern Pacific launched a ten-year Dieselization program beginning on October 12, 1946, with an order for 20 Diesel locomotives of 6,000 horsepower each." (When delivered, Southern Pacific's F units were painted in its now

Looks can be deceiving: Southern Railway 4139, seen in Washington, D.C., would appear to be an F3A. It was built as an F3 but later upgraded to an F7. (photo by Jim Shaughnessy)

Freight power: Southern Railway 4170 is seen alongside a Burlington 2-8-2 Mikado at Centralia, Illinois, on August 31, 1957. Upgrading an F3 to F7 involves internal electrical improvements. (photo by Jim Shaughnessy)

An advantage of diesel-electric power is flexibility — units can be mixed and matched as needed. New York Central F units and a GP7 climb with westward freight at Washington, Massachusetts, on June 15, 1964. (photo by Jim Shaughnessy)

Southern Pacific had the largest roster of postwar F units. Hundreds of F3s, F7s, and FP7s prowled its mountainous empire. SP F7A 6249 leads a freight at Azalea, California, on the Shasta Route in the early 1960s. (photo by Bob Morris)

Their dynamic brake grids and fans whining loudly, four SP F7s drop downgrade at Shasta Retreat, minutes away from a crew change at Dunsmuir, California. (photo by Bob Morris)

SP's F units were originally painted in its now-famous "Black Widow" livery — black body with silver noses, accented with orange and red striping. Black Widow F7 6387 leads a freight at Sawmill Curve, upgrade from Dunsmuir, California. (photo by Bob Morris)

Southern Pacific F7A 6321 rests on a frosty evening on the west slope of Donner Pass at Truckee, California. SP used a dual-headlight arrangement — top position was an oscillating Mars light, bottom was a fixed light. (photo by Bob Morris)

In 1963, SP 6266 ascends the summit of Donner Pass at Norden, California. SP maintained miles of wooden snow sheds to protect this important transcontinental crossing from deep, drifting snow. (photo by Bob Morris)

The roar of 16-567Bs in run-8 and wafts of black under-aspirated exhaust proceed from the exit of SP extra 6277 West from Tunnel 41 at the Summit of Donner Pass. Five F7s — 7,500 hp — lead this freight, followed by additional F7s shoving hard against the caboose. This was the drama of heavy-mountain railroading in the first-generation diesel era. (photos by Bob Morris)

famous "Black Widow" livery, a black body with silver front and red and orange nose striping.)

In many ways, the new F unit closely resembled the FT. Among the most important internal changes was a new engine design, model 16-567B, which was rated 1,500 hp, making it 150 hp more powerful than older 16-cylinder 567 engines.

On the electrical side, Electro-Motive's significant innovation was the D-12 combined generator and alternator. In a diesel-electric it is necessary for the electrical equipment to match the output of the engine. However, this presented production problems with the new F. Although the demonstrator was a success, and Electro-Motive moved forward with publicity for the new model, the company was not ready for mass production of the D-12 generator. To meet road diesel orders, it built an interim F2 model that incorporated many of the features advertised by the postwar demonstrator, including the new engine and improved carbody, but employed a variation of the older D-8 generator and was consequentially rated at just 1,350 hp — same as the FT. When the new generator was finally ready, the new line debuted the designed model F3. Production F2s were built for just five months, between July and November 1946, and totaled just 104 units — 74 "As" and 30 "Bs."

Externally, the F2s and earliest F3s (described as phase I, see explanation below) appeared nearly identical. Electro-Motive had lengthened the F-unit carbody to make more room and ease maintenance. The A units measured 50 ft. - 8-3/8 in long, while the B units were 50 ft. - 1/4 in. long. Based on specifications for the Boston & Maine F2s, the carbodies were 14 ft. - 1/2 in. tall, and the distance from the rail to the top of the horn was 14 ft. - 11-1/4 in. All the F2As featured small number-board and class-light combinations on each side of the nose like those on the FTAs, also a feature of the early F3As. Later F3As and subsequent F-unit A models used a larger, protruding rectangular number board and separate rounded class lamp. The sides featured a row of three port-hole windows, spaced much further apart than the sequence of four portholes on the FTs. The center window was hinged for access. On top of the locomotive was a row of four shrouded

Santa Fe had separate fleets of freight and passenger F units. An A-B-B set of F7s lead high-level streamlined cars at Pinole, California, in 1967. (photo by Bob Morris)

A pair of maroon and cream Soo Line Fs lead a train of timber products past a local mixed train at Trout Lake, Michigan, in 1961. (photo by Richard Jay Solomon)

radiator cooling fans protruding several inches above the carbody.

David P. Morgan noted in the September 1949 issue of TRAINS Magazine that the first railroad to purchase the F2 was the Atlantic & East Carolina, giving this railroad the distinction of being the first short line to purchase Electro-Motive road diesels. The line bought two units numbered 400 and 401. It operated a 94-mile main line between Goldsboro and Morehead City, North Carolina. According to March/April

The F unit is part of American railroad culture. Monon F3As are pictured on this 1950s painted postcard of The Hoosier leaving Chicago. (Richard Jay Solomon collection)

1996 *Diesel Era* magazine, Atlantic & East Carolina assigned its F2s to its daily passenger trains. However, unlike many later passenger Fs, these were not equipped with boilers for train heat. The units need to conform to very light axle weights in order to avoid damaging the railroad's lightly built track. They weighed just 226,790 lbs., making them seven tons lighter than other F2s.

Most F2 buyers — Atlantic Coast Line, Boston & Maine, Burlington, Southern Railway affiliate Cincinnati, New Orleans & Texas Pacific, Minneapolis & St. Louis, New York Central, and Rock Island — were already FT owners. Typically these companies bought F2s to increase fleet flexibility. Why assign a 5,400-hp A-B-B-A FT set to a freight if three units could haul the tonnage effectively?

Boston & Maine, for example, bought two fleets of F2s. Initially, it ordered 15 A units to pair with its existing FT A-B sets, allowing it to run A-B-A sets in freight service, thus overcoming some of the limitations fixed A-B pairs. Later it bought three A-B F2 sets with steam generators for passenger service. Both types used the standard 62:15 gear ratio, 40-in. wheels, and D7K1 traction motors. Based on specifications for the three A-B sets, the A units weighed 238,030 lbs., and the A-B pair could produce 65,000 lbs. of continuous tractive effort at 13 mph.

Another notable buyer of the F2 was National Railways of Mexico (NdeM), which bought 28 units (14 As and 14 Bs), and later became a large customer of F-unit models. Mexico's F2 fleet was the largest and also the longest surviving. In the 1960s, most American F2s were either traded back to Electro-Motive as credit for newer units, typically high-horsepower four-motor "GPs," or sent for scrap. The NdeM units largely survived into the 1970s. March/April 1996 *Diesel Era* credits this to rebuilding into "F7AMs/F7BMs" by Electro-Motive during 1955 and 1956, which included the replacement of original engines with the significantly more reliable 16-567BC design. Externally, rebuilding altered the locomotives by removing the characteristic center porthole and adding side louvers in the arrangement typical of F9s.

Chicago & North Western train No. 501 under gray skies at Madison, Wisconsin, on January 31, 1957. Leading is C&NW F3A No. 4054A, a late-era F3 in the "phase IV" carbody. (photo by John Gruber)

F3 Sets a New Standard

As explained above, the F2 was just a prelude to the F3, which finally began production at the end of 1946. It was the F3 that rapidly became the best-selling locomotive on the North American market. It outpaced competing Alco and Baldwin models, and its production totals outsold even Electro-Motive's own FT. General Motors' C.R. Osborn noted in the October 1946 *Railway Age* that prior to regular production, 30 different railroads had placed orders for the F3 based on the demonstrator's performance. During the F3's comparatively short production run, which ended in February 1949, a combined total of 1,807 (A and B) units were sold for use on North American railroads.

What was so special about the F3? Electro-Motive successfully addressed performance and reliability problems identified with its FT. In short, the F3 was a more powerful, more versatile, and more reliable machine than the FT, and easier to maintain too.

The nominal increase in output was one selling point. A four-unit F3 was rated at 6,000 hp — 600 more than a four-unit FT set. The F-units' output matched diesel competition from Alco, and later Baldwin and Fairbanks-Morse. The F3 was also a versatile locomotive. Electro-Motive had dispensed with the permanent connections between A and B units, thus allowing railroads to order and operate F units in any combination they desired. This allowed railroads to adjust consists as needed for regular service.

Traditionally, most American railroads custom ordered steam locomotives for specific applications. When considering its motive power needs, railroads might have ordered a dozen Pacifics with 72-inch driving wheels for express passenger service, two dozen Mikados for general freight service, and ten low-drivered Santa Fe types for drag service. In addition, it might have separate types for helper service, local passenger service,

Milwaukee Road continued to operate a large fleet of F units through the 1970s, often assigning them to secondary duties. A pair of tired-looking F7As work Madison, Wisconsin, on March 29, 1978. (photo by John Gruber)

On November 18, 1977, a Milwaukee Road F7 was photographed against the backdrop of the Wisconsin state capitol at Madison. Many of Milwaukee's Fs have since been preserved. (photo by John Gruber)

and so on down the line. Locomotives would often be specifically tailored for a type of service on a specific district. Weight on the drivers, size of the drivers, cylinder size, boiler capacity, tender capacity, maximum speed, and hauling capacity all dictated the design of a locomotive. By the 1920s, railroads were moving toward steam locomotives for more general-purpose applications. Canadian National had ordered dozens of modern 4-8-4s with relatively low-axle loading for general service. So when General Motors advertised that one locomotive could be adapted for many types of service, it really struck home. Why order a dozen different kinds of engines if just one model would do? General Motors promoted the F3 as an adaptable general-purpose locomotive and advertised in June 1947 *Railway Mechanical Engineer*:

"For years, railroads have needed a locomotive which, because of its versatility and range, could serve successfully as either a heavy-duty fast passenger locomotive, freight locomotive, or a combination of both.

"That wide-range locomotive is now here — the General Motors F3 — performance-proven by tests and actual service on 31 important American railroads.

"Witness the job two of these F3 Diesels are doing in combination passenger-freight service on Gulf Mobile & Ohio Railroad between Venice, Ill., and Kansas City, Mo.

"An F3 leaves Venice on a freight train, No. 90, about 12:40 P.M. and arrives Roodhouse, Ill., about 5:00 P.M.

"On arrival at Roodhouse, the locomotive is cut off the train and put on a passenger train, No. 23, leaving Roodhouse [at] 12:35 A.M., arriving [at] Kansas City 7:40 A.M., a distance of 251 miles.

"On the return trip, it leaves Kansas City [at] 11:00 P.M., arriving Roodhouse [at] 6:00 A.M., where freight train No. 91 is called. The F3 is taken off the passenger train and coupled onto the freight about 6:15 A.M. It arrives at Venice, according to [the] freight schedule about 8:10 A.M.

In a rare act of corporate nostalgia, Western Pacific shopped and repainted F7A No. 913 into its old orange and silver scheme in April 1978. A year later it leads train SJT over California's Altamont Pass. (photo by Brian Jennison)

Wabash's Detroit, Michigan, to Niagara Falls and Buffalo, New York, mainline operated across Ontario. As a result, Wabash had a fleet of Canadian-built F7As. On August 1, 1957, working solo, No. 1163 leads a freight at Welland Junction, Ontario. (photo by Jim Shaughnessy)

"Immediately on arrival at Venice, the Diesel is placed on a return freight for Roodhouse, and the cycle is repeated.

"The two General Motors Diesel locomotives alternate on this punishing daily grind, averaging 410 miles a day on the combination passenger-freight runs."

It was offered with eight different gear ratios ranging from 65:12 for slow-speed drag service to 56:21 for fast passenger service. The first number indicates the number of teeth on the bull (axle) gear, and the second number indicates the teeth on the pinion gear. At 65:12 the locomotive could deliver 42,500 lbs. continuous tractive effort at about 11 mph and operate at an advertised top speed of 50 mph. By contrast, the 56:21 ratio delivered 21,000 lbs. continuous tractive effort at about 18 mph while allowing a top speed of 102 mph. (Both figures are based on 25 percent adhesion.) Railroads would typically order locomotives with gearing as close to the type of service intended as possible. A locomotive with high-speed gearing for

Wabash bought F7s to replace its 2-8-2 Mikados in freight service. No. 1160 rests at Ft. Erie, Ontario, as an aged Canadian National Mikado works the yard. (photo by Jim Shaughnessy)

POSTWAR Fs

Wabash's F7A No. 1189 has been preserved and restored by the Monticello Railway Museum at Monticello, Illinois. (photo by Brian Solomon)

Close-up view of the flush angled pilot on Wabash No. 1189. This type of pilot was more commonly used on passenger service units than on freight Fs. (photo by Brian Solomon)

passenger service could be assigned to freight work, but would be greatly limited when working in heavily graded territory.

Figures published in Electro-Motive's F3 Operating Manual illustrate the difficulties of using passenger-geared Fs in freight service on a grade. A four-unit set of F3s rated at 6,000 hp with 62:15 gearing, by far the most common arrangement, is designed for a maximum speed of 65 mph and rated to lift 4,780 tons up a 1-percent grade. On a 2-percent grade, this same set is only rated for 2,460 tons. By contrast, a four-unit F3 set with 59:18 gearing designed for 83 mph top speed can only lift 3,700 tons up a 1-percent grade, and just 1,870 tons up a 2-percent grade.

If a railroad wanted to change the type of service to which a locomotive was assigned, the gear ratio could be changed at the railroad's shops. Rebuilding by the manufacturer or substantial modifications to the locomotive were not required.

An April 1947 General Motors advertisement boasted:

"28 Locomotives in one! The Wide Range F3.

"The General Motors F3 matches the locomotive to the job. Its range is so broad that with proper gear ratio it can be used to drag heavy-tonnage freights or take over heavy-duty fast passenger schedules.

"By a simple change of one of seven standard gear ratios, the F3 can be equipped to perform over a range from that of the heaviest dragging freight locomotive, for mountain terrain, through that of a combination freight and passenger locomotive, on up to that of a heavy-duty passenger locomotive capable of pulling long, standard-weight Pullman trains at speeds in excess of 100 miles per hour. It will negotiate most mountain grades without a helper."

General Motors goes on to describe how the F3 may be singly used at 1,500 hp or in combinations up to four units, giving railroads multiples of 1,500 hp to work with. How did GM justify "28 locomotives in one"?

"Multiply the four combinations of units available by the seven gear ratios and you get twenty-eight different locomotives possible in the one basic locomotive!"

General Motors Diesel Limited builder's plate on Wabash No. 1189. GMDL built Electro-Motive locomotives for Canadian users. (photo by Brian Solomon)

Later F7s and most F9s use this style of stainless-steel radiator intake grill, characterized by three rows of vertical slits. (photo by Brian Solomon)

Externally, a drag-service F3 would appear the same as one geared for fast passenger service.

After 1947, the Interstate Commerce Commission introduced new and more restrictive regulations regarding maximum train speeds. Unless a railroad employed advanced signaling systems, such as automatic train stop or cab signals, top speed for passenger trains was essentially limited to 79 mph, and freight to 70 mph. As a result, railroads had limited applications for the variety of higher gear ratios offered on the F3.

Some railroads embraced General Motors' philosophy and ordered hundreds of F-units, assigning them to a wide range of tasks. Other lines sampled diesels similar to the way they had steam, buying a few of each model from all the major builders, and assigning each type to the most appropriate location.

The F3s featured much improved reliability. General Motors' advertising had emphasized the FT's high availability; what the advertising didn't mention was how those high availability figures were attained. Traveling diesel maintainers were some of the great unsung heroes of the early diesel era. To ensure FTs were in good running condition and they got over the road in working order, many railroads assigned specially trained diesel maintainers to ride along with FTs. Part of the maintainers' responsibilities were to routinely inspect locomotive equipment, looking for signs of wear or pending failure.

An article in the May 1946 *Diesel Railway Traction* explains that the maintainer would note lube oil pressure, inspect fuel injection equipment, the engine governor, contractors, engine crankcase, and valves, while looking for water and oil leaks. The maintainer would also routinely implement small repairs to locomotives while they were moving. On occasion, more intensive repairs might be required on the road. The arrangement of the engine block in the FT carbody had been designed to allow for the removal of cylinder liners

In later years, most railroads added grab irons to F units to allow easier access to the top of the nose and roof. (photo by Brian Solomon)

Wabash merged with Norfolk & Western in 1964, and its F7As were renumbered and later repainted for their new owner. N&W No. 3726 seen at Buffalo, New York, was former Wabash No. 1189A, sister locomotive to the unit preserved at Monticello, Illinois. (photo by Doug Eisele)

Norfolk & Western leased some of its former Wabash F7As to the Central Railroad of New Jersey in the early 1970s. N&W No. 3689 rests between runs at Bethlehem, Pennsylvania. (photo by George Kowanski)

from the engine without opening hatches or otherwise exposing the locomotive. *Diesel Railway Traction* reported that Rio Grande carried out approximately 40 percent of normal engine maintenance on the road. While this enabled the railroad to claim extraordinarily high unit availability, it also indicated that the FTs were maintenance-intensive machines. With its F3, Electro-Motive emphasized improved dependability, operating the locomotive for 60,000 miles without a major shop visit. New technology was also introduced. Controlling 6,000 hp seemed intimidating to some railroaders. In the early days of FT operations, locomotive engineers used to operating steam power were known to have pulled too hard too quickly, breaking coupler knuckles and pulling draw bars. With a variety of gear ratios available and even higher horsepower, Electro-Motive needed to allay fears that the new locomotive would destroy equipment or otherwise be difficult to operate.

To address questions of adequate power control, General Motors explains in the August 1947 *Railway Mechanical Engineer* advertisement entitled, "The Locomotive with the Mechanical Brain":

"The new engine governor of the F3 gives maximum speed to fit any type of train service... This highly selective control unit responds directly to the engineman's throttle. Yet because of its exact timing action, a rough start is practically impossible, no matter how fast the hand throttle is pulled.

"Speed and load make no difference. Heavy freights can be started with practically no thought of broken coupler knuckles. Frequent-stop passen-

Burlington Northern inherited large numbers of F units from its predecessors. In the mid-1970s it invested in overhauling and operating them in road service until the early 1980s. On July 21, 1978, BN train No. 148 is led by 830, a former Northern Pacific F9A. (photo by Tom Carver)

The last BN F unit to be repainted was No. 724, seen on April 29, 1980, on the turntable at Auburn, Washington. This was one of many former Northern Pacific F7s that worked for BN. (photo by Tom Carver)

ger trains can be handled with the fast pick up needed.

"Engine and generator are perfectly balanced — power output and speed completely synchronized. There's no drop in the performance curve. For each throttle position, a constant horsepower output is maintained."

Other reliability issues were addressed too. The longer carbody eased maintenance. A balanced design replaced the asymmetrical wheelbase featured on the FT. More significant was the complete redesign of auxiliary appliances — radiator fans, traction motor blowers, etc. In place of mechanically driven systems used on the FT, which required a network of belts and pulleys, F3 appliances were electrically driven with three-phase AC motors. The alternator portion of the D12 generator provided power. On the FT radiator, shutters had been manually operated and required regular attention from the fireman. Failure to operate the shutter properly could result in an overheated engine. With the F3 carbody design, the engine-room arrangement was totally reconfigured. Radiators were relocated from the ends of the engine compartment to the center of the locomotive and positioned at diagonals above the engine block. Instead of belt-driven, roof-mounted radiator cooling fans at the ends of the engine compartment, a row of four electrically driven radiator fans centered over the engine were used to automatically cool the engine. Belt-driven fans made such an arrangement impractical. Electrical operation had additional advantages. Using sensors, the electrical control circuits were wired to switch the roof fans on and off as needed to keep engine coolant temperatures within desired tolerances.

The nose of BN No. 732, a former Northern Pacific F7A, reveals the signs of 30 years of hard service. It was built in 1951 and photographed in 1981. Recession and new locomotives would spell the end of the line for these old machines. (photo by Tom Carver)

Moving the radiators allowed for relocation of dynamic brake grids and blowers. These were placed within the carbody ahead of the diesel engine and directly behind the cab. There were two sets of grids and blowers. Hot air was exhausted through a pair of long vents at the top of the locomotive above the grids. This arrangement simplified maintenance and made for easier inspections of the equipment.

Electrically operated traction motor blowers allowed for more effective positioning and better exhaust. The FT suffered from poor traction-motor ventilation, and the tendency for heat to travel through the locomotive to the trailing units. In the new arrangement, blower exhaust was directed up and out of the locomotive.

Coal-hauling Chesapeake & Ohio began dieselization of its road freights later than most railroads. Its first road units were F7s ordered in September 1950. C&O F7A No. 7069 in faded factory paint is seen at Buffalo, New York, in August 1968. (Doug Eisele collection)

During the course of F3 production, Electro-Motive engineers implemented additional improvements. As noted, the number boards were made larger and more prominent. The air-intake vents were redesigned — the older external "chicken wire" covering was altered, then replaced altogether by more attractive stainless-steel coverings with long lateral vents. David P. Morgan, writing in the September 1949 issue of TRAINS, suggested that for the purpose of Electro-Motive's engineering department "an interim F-5 model was created." The F5 was not an official designation, but was used to describe late-production F3s (built from the end of August 1948 to February 1949) that incorporated some improvements made standard with the F7 line.

The Most Popular F

Electro-Motive's extremely popular F3 model was succeeded in February 1949 by Electro-Motive's latest F unit, the F7. Although externally the F7 appeared virtually identical to late-model F3s, this new model incorporated a

Three-phase alternating-current motors powered the auxiliary appliances. This required less maintenance than direct-current motors, and was substantially more reliable than belt-driven appliances. The F3's D12 main generator was crankshaft driven and featured a built-in 149-volt three-phase alternator to power AC appliances. In addition, a 10 kW auxiliary generator was used to supply current for control of circuitry and lighting, and to charge batteries.

Other design improvements included a better airbrake schedule fully compatible with early brake controls, a new transition meter that made it easier to calculate short-time traction motor ratings, and a higher-capacity steam generator.

The new 16-567B engine was more powerful, but also significantly more reliable than earlier 567 models. Electro-Motive's earlier 567A is a notoriously leaky engine. Its flawed cooling circuit in the cylinder liners was prone to letting excessive amounts of water into the lube oil. If left unattended, water in the lube may lead to a multitude of problems that damage the engine, causing road failures and costly repairs. Other improvements included a new governor design and improved synchronization between the engine and generator that allowed for more balanced and efficient operation.

Reading F3B No. 260B and GP7 No. 632 leads a freight on the Central Railroad of New Jersey at Communipaw, New Jersey, on April 4, 1958. (photo by Richard Jay Solomon)

In 1962, US Steel constructed a new 78-mile line from the Union Pacific at Winton Junction to an iron mine at Atlantic City, Wyoming. Former Bessemer & Lake Erie F7s worked the line in heavy service. Six F7s lead an ore train at Winton Junction on September 3, 1973. (Doug Eisele collection)

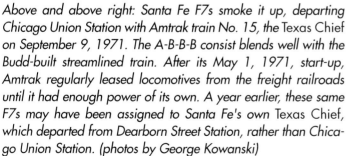

Above and above right: Santa Fe F7s smoke it up, departing Chicago Union Station with Amtrak train No. 15, the *Texas Chief* on September 9, 1971. The A-B-B-B consist blends well with the Budd-built streamlined train. After its May 1, 1971, start-up, Amtrak regularly leased locomotives from the freight railroads until it had enough power of its own. A year earlier, these same F7s may have been assigned to Santa Fe's own *Texas Chief*, which departed from Dearborn Street Station, rather than Chicago Union Station. (photos by George Kowanski)

Right: Santa Fe F7s lead the *Grand Canyon* at Chicago in the early 1960s. Santa Fe preferred Fs over Es for most passenger services. (photo by Richard Jay Solomon)

POSTWAR Fs

Santa Fe F3B No. 347B and F7A No. 347C are preserved at the California State Railroad Museum in Sacramento.

Santa Fe's passenger Fs had stainless-steel side panels to match the railroad's Budd-built stainless-steel passenger cars. The famous "Warbonnet" paint livery was designed for the railroad by Electro-Motive's Leland A Knickerbocker. (photos by Brian Jennison)

In the 1970s, Santa Fe experimented with paint liveries on its F units. F7A No. 329 wears a freight version of the traditional Warbonnet. (photographer unknown, Doug Eisele collection)

number of technological improvements in both performance and reliability. Now symbolic of the early diesel era, the F7 soon became one of the best-selling diesels of all time. The F7A is undoubtedly one of the most recognizable and most familiar of all locomotive designs. It has been the prototype for innumerable model-makers and has often been featured in advertising and period literature. More important than good looks, the F7 achieved new standards for reliably and performance, making it the envy of the industry. During its four-year production run, America's railroads were buying diesels as fast as production allowed. Railroads were frantically trying to replace steam. The F7 was by far the most desired locomotive design in North America. According to Louis Marre's *Diesel Locomotives: The First 50 Years*, 2,366 A units and 1,483 B units were built for service in the United States, Canada, and Mexico. In addition, 371 FP7s (all A units) were built. Electro-Motive's total F7 production greatly exceeded the combined production total of all four-motor (B-B wheel arrangement) carbody locomo-

In September 1981, dead F units were lined up at Auburn, Washington. After 30-plus years of service the end of the road had finally come. (photo by Tom Carver)

tives built by Alco, Baldwin, and F-M at that time. Furthermore, with its superior products, including F7, Electro-Motive effectively drove the weakest manufacturers out of the business.

F7, the Better F

Primary changes with the F7 involved much improved traction-motor design, expanded-range dynamic brakes, better fuel injectors, and the adoption of automatic motor-transition as a standard feature. Other improvements included better insulation and engine room airflow, better boiler piping, and a more efficient steam generator, all of which greatly improved cold weather operation. As mentioned earlier, notable external

Soo Line F7A No. 2201A and GP30 No. 711 lead a freight across the Mississippi River at Minneapolis, Minnesota, in 1974. (photographer unknown, Doug Eisele collection)

POSTWAR Fs   53

Accident victim: Soo Line F7A No. 2228A had sheet-metal surgery on its nose-section. The cylindrical headlight housing was removed and patched at the railroad's Shoreham Shops in December 1973. (photographer unknown, Doug Eisele collection)

features were shared with late-model F3s, but basic external dimensions remained unchanged.

Electro-Motive's F7 was introduced simultaneously with the FP7 and E8, the latter two types aimed at the passenger train market. All three models incorporated the same technological improvements. With the introduction of the new models, older models were superseded and discontinued.

Improved Performance and Reliability

The upgrade from FT to F3 boosted engine output, while the upgrade from F3 to F7 boosted continuous tractive effort ratings. Like the F3, the F7 was rated at 1,500 hp per unit. However, the F7 used D27 traction motors in place of the older D17 motors. Consistent with Electro-Motive's design philosophy, the new motors conformed to the size specifications of the older models, so the D27, D17, and even the older D7 were interchangeable. The key ingredient that made the D27 a better motor was significant improvement to insulation materials used on the armatures and field coils. Older motors had used organic components in insulation, which made the motors more susceptible to damage from overheating. To prevent damaging motors, Electro-Motive had strict short-time overload ratings, which limited the ability of diesels to start very heavy trains and operate at slow speeds under full load. The May 1949 *Railway Mechanical Engineer* offers a detailed description of the D27 motor improvements. Inorganic insulation consisted of a silicon, glass, and mica compound, which greatly improved thermal conductivity, and thus reduced the damaging effects of heat on the motor. Bolted and soldered stator connections were replaced by brazed connections. Heavier cables were used in traction-motor circuits, and traction-motor cooling capacity was improved. From an operating standpoint, this raised the continuous motor rating from 700 to 825 amps, and it dramatically improved the length of time the motors could be overloaded without risk of damage.

In order to start trains and keep them moving on heavy grades, diesel electrics are operated with overloaded traction motors for limited time. The amount of

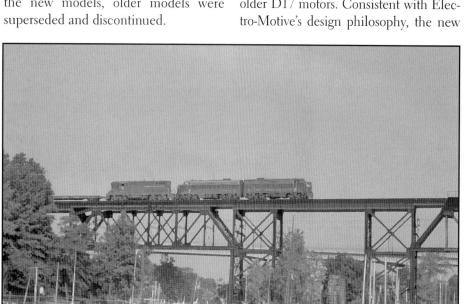

On August 26, 1973, Norfolk & Western F7A 3657 leads a freight across the Rocky River Bridge at Lakewood, Ohio. N&W never bought new F units, and all Fs on its roster were inherited from the Wabash in 1964. (photo by James P. Marcus. Doug Eisele collection)

These views of Erie-Lackawanna F7B No. 6112 at Hornell, New York, show it first in faded paint prior to repainting on September 14, 1975, and precisely six months later undergoing repainting at the company shops. In this case, the maroon stripes on the distinctive Erie-Lackawanna paint livery resulted in 6112's whole body painted maroon with gray and yellow painted over later. (both photos by Bill Dechau, Doug Eisele collection)

up a 1-percent grade than an F3, while in short-time ratings it could haul 30 percent more tonnage. The locomotive's improved hauling ability needed to be matched with better braking. So the F7 was designed with significantly higher dynamic braking capacity.

Automatic transition, which had been offered as an option on F3, was standard on F7. With manual transition, responsibility to make motor transitions at the correct times was entirely the responsibility of the engineer. While allowing for individual control, this could result in damaged electrical equipment if transitions were not initiated at appropriate times.

Its improved fuel injection system made the F7 a more fuel-efficient locomotive. April 1949 *Diesel Railway Traction* explained that using larger plungers and bushings in combination with shorter fuel injection timing enabled the use of lower grades of diesel fuel and thus increased fuel efficiency.

the overload determines the length of time motors may be overloaded. This is carefully gauged by published "short-time" ratings. When the limits of short-time ratings are reached, the locomotive must be shut down and its motors allowed to cool. Failure to do this damages the motors. Lugging up a heavy grade, if a locomotive reaches the limits of short-time ratings, it must be shut down. This had been a perceived drawback to early Electro-Motive diesels, as early Alco and Baldwin locomotives had a superior ability hauling heavy trains up steep gradients.

Improved traction motors equaled more pulling power. Depending on the gear ratio, the F7 delivered 21 to 24 percent more tractive effort than the F3 using the same 16-567B engine and D12 main generator. Using the most common 62:15 gear ratio as an example (where the F3 was rated at 32,500 lbs. continuous tractive effort), the F7 was rated at 40,000 lbs., an increase of 23 percent. According to an article in the April 1949 *Diesel Railway Traction*, Electro-Motive claimed that in freight service its F7 could lift 25 percent more tonnage

On July 4, 1971, Erie-Lackawanna F7A No. 6114 is in the process of being repainted at the Hornell, New York, shops. This locomotive was built in July 1949 as Lackawanna No. 611C. (photo by Doug Eisele)

POSTWAR Fs   55

Erie-Lackawanna F7A No. 6341, former Lackawanna No. 634A, sits under stormy skies at Hornell, New York on September 21, 1975. Erie-Lackawanna's F-unit fleet was scrapped shortly after Conrail assumed operations of the bankrupt carrier in April 1976. (photo by Bill Dechau, Doug Eisele collection)

Model FP7

To accommodate the need for long passenger runs, Electro-Motive offered a variation of the F7A, designated FP7 (in some period literature this was also designated FP7A, but since the locomotive was only built as an A unit, FP7A seems redundant). It had all the features of the F7A and was four feet longer to allow room for an additional 950-gallon water tank for steam heat. This provided the FP7 a total water capacity of 1,750 gallons — almost double the F3A capacity. Greater capacity allowed an FP7 to operate significantly longer distances between water stops than earlier F units. Visually comparing an F7 and FP7, the extra length is evident between the porthole window beyond the back of the cab and the first set of air-intake louvers (below the long stainless-steel air-intake cover).

Although by no means as common as the F7, 372 FP7s were built for use in North America. The list of railroads that owned and operated FP7s is impressive and includes: Alaska Railroad, Atlantic Coast Line, Atlantic & West Point Rail Road, Canadian Pacific, Chesapeake & Ohio, Chicago & Eastern Illinois, Chicago & North Western (via Chicago Great Western), Chicago Great Western, Clinchfield, Florida East Coast, Frisco, Georgia Railroad, Louisville & Nashville, Missouri-Kansas-Texas, Missouri Pacific, National Railways of Mexico, Ontario Northland, Pennsylvania Railroad, Reading, Rock Island, Soo Line, Southern Railway (through its Cincinnati, New Orleans & Texas Pacific affiliate), Southern Pacific, Union Pacific, Western Pacific, and Western Railway of Alabama. In addition to the lines already listed, a number of lines operated secondhand FP7s, such as Amtrak, which inherited former Southern Pacific FP7s after its creation in 1971.

Years of demanding service took their tool on F-unit carbodies. Notice the plywood panel on 7101 and the missing radiator grill on the following unit. These two Fs are working as helpers at Susquehanna, Pennsylvania, in October 1972. (photo by George C. Corey)

Although intended for use on steam-heated passenger services, as railroads curtailed passenger operations during the 1950s and 1960s some companies routinely assigned FP7s to freight work. Take the Clinchfield, for example. When this Appalachian coal-hauler bought its sole FP7 in 1952, it assigned it to the line's one remaining passenger ser-vice. This tri-weekly train between Elkhorn City, Kentucky, and Spartanburg, South Carolina, was discontinued at the end of April 1955. Clinchfield reassigned the FP7 to freight service, joining F3s and F7s. Interestingly, this lone locomotive, originally number 200, survived the wave of corporate mergers that relegated Clinchfield and its successors to history. Numbered 118 in later years, it ended up serving as one of CSX's executive F units and was used to haul CSX passenger specials until its retirement in 1993.

F9

The F9 was Electro-Motive's final freight and passenger F-unit model, and it was in production from early 1954 until 1960. It was introduced along with ten other locomotive models, most of which also used the "9" suffix (GP9, E9, and etc), and each of which also superseded older models (GP9 taking the place of GP7, E9 the place of E8, etc.) The most important feature of these new models was Electro-Motive's much improved engine, designated 567C.

In the fading light of July 16, 1970, Baltimore & Ohio F7A No. 4499 hits a road crossing on the old Buffalo, Rochester & Pittsburgh 4th Subdivision in Rochester, New York. (photo by Doug Eisele)

In the late 1960s and early 1970s, financially strapped Central Railroad of New Jersey leased F7s from N&W and B&O. CNJ No. 15 was renumbered from B&O 941, built in 1950. Paint is B&O's with CNJ "Statue of Liberty" insignia. (photo by George Kowanski)

POSTWAR Fs

On September 4, 1972, the locomotive line up at Sayre, Pennsylvania, included Alco C-420 No. 412, C-628 No. 627, and F7A No. 562. Lehigh Valley had bought FTs in 1945, and followed up after the war with F3s and F7s. (photo by Doug Eisele)

These former Chicago & North Western F7s were painted in the Lehigh Valley livery for excursion work and are seen on a fan trip at Kingsland, New Jersey, on May 20, 1995. LV No. 576 was originally C&NW No. 4073A, later No. 418; LV No. 578 was No. 4087C, later No. 420. Both locomotives worked on NJ Transit after leaving C&NW. (photo by Patrick Yough)

Chicago & North Western 4075A shoves on the back of a suburban push-pull set on September 8, 1971. C&NW was one of the first railroads to use diesels in push-pull service. (photo by George Kowanski)

Rock Island F7A 677 and one of the railroad's unusual AB units lead a suburban train in Chicago on September 10, 1971. Derived from an E unit, the AB was a 1,000 hp flat-front diesel that also contained a baggage car designed for the Rocky Mountain Rocket. (photo by George Kowanski)

This was the culmination of evolutionary advancements from more than 15 years of experience with the 567 block. It was more powerful and substantially more reliable than earlier 567s. Cylinder bore and stroke remained unchanged, but Electro-Motive increased maximum rotational speed from 800 to 835 rpm in order to boost output. The engine was offered in 6-, 8-, 12-, and 16-cylinder versions. The 16-567C was used by the F9, which boosted F-unit output by 250 hp. The F9 was at 1,750 hp, and a four-unit sat at 7,000 hp. The engine crankcase had been redesigned to withstand greater stress. Cylinder liners were redesigned with an improved cooling circuit using O-ring style inlet manifolds on both top and bottom. This replaced the top water seals used on older 567s, which suffered from excessive leakage. The new arrangement also provided better cylinder head cooling.

In addition to the 567C engine design, Electro-Motive revamped the electrical system. The D37 traction motor was introduced, superseding the D27 motor. According to January 1954 *Railway Locomotive and Cars*, the D37 used an improved molded coil to better seal out moisture while permitting an increase in traction motor ratings and a virtual elimination of arbitrary short-time overload ratings. Improved synthetic insulation allowed for greatly improved motor reliability. Using Teflon in place of mica, insulation lowered friction. The D37s produced higher output while requiring substantially less maintenance. Advances in control circuitry allowed for a simplified and totally enclosed locomotive electrical interlocking machine that required less space and improved reliability.

The F9 specifications as published by Electro-Motive in its *Operating Manual F9 Locomotives* are as follows: The F9A is 50 ft. 8 in. long; the F9B 50 ft. 0 in.; both are 15 ft. 0 in. tall and 10 ft. 8 in. wide. Standard wheel diameter is 40 in., both the F9A and F9B each weigh approximately 230,000 lbs. fully loaded — each locomotive unit carries 1,200 gal. of fuel oil, 200 gal. of lubricating oil, and the F9A has 230 gal. of cooling water capacity.

As with the earlier F units, the F9 was available with a range of gear ratios. The lowest was 65:12 with a 55-mph maximum speed and the highest was 56:21 with a 105 mph maximum.

Western Maryland F7A 235 wears the railroad's final paint livery. Most Fs wore the more conservative black with gold striping. (photo by Doug Eisele)

The unmistakable profile of an F unit: Penn Central No. 1784 crawls across a grade crossing in Rochester, New York, on July 9, 1973. This former New York Central F7A was one of hundreds of F-prowling PC rails in the early 1970s. (photo by R.R. Richardson, Doug Eisele collection)

Higher maximum speeds were possible as a result of improved performance characteristics.

Electro-Motive was confident of the F9's pulling power, and its operating manual doesn't even list approximate tonnage ratings. Instead, it says under section 213, "Locomotive Operation At Very Slow Speeds":

"The operation of an F9 locomotive, regardless of gear ratio, is not governed by any specific short-time ratings.

"In most cases, the locomotive may be operated up to the limit of the adhesion attainable.

"F9 locomotives pulling tonnage trains at very slow speeds should be operated with the throttle in Run 8 position. In the event of a wheel slip indication (wheel slip light flashes on), the locomotive wheel slip control system will automatically

Penn-Central FP 7 No. 4369 is former Pennsylvania Railroad 9869. On April 7, 1975, it works an ore train at Freeport, Pennsylvania. (photo by Jim Marcus, Doug Eisele collection)

apply sand to the rails (automatic sanding feature "cut in") and reduce the power to a point were slipping stops. If continuous wheel slipping on sand occurs due to unusual rail operating conditions, the throttle can be reduced. Under these circumstances, the F9 locomotive can operate at reduced throttle, provided it is not necessary to reduce below the 5th throttle position to correct for a continuous wheel slip."

Despite this advice, since the F9 still used conventional direct-current traction motors, it was possible to overload them, although the F9 was more durable than all earlier models.

Without question, the F9 was the most refined F-unit model — it was the most powerful, most reliable, and easiest to maintain of the traditional freight and passenger Fs (technically, the last 30 FL9 "hybrids" were the most powerful Fs; these were rated at 1,800 hp – see Chapter 4). Yet, it was by no means the best-selling F; to the contrary, its sales were relatively poor. Compared to the 1,111 F3As, 695 F3Bs, 2,366 F7As, and 1,483 F7Bs sold between 1945 and 1953, the F9A sold just 87 units, and the F9B only 154 units in North America. While the FP9 passenger version was nearly exclusively built for railroads in Canada, Mexico, as well as Saudi Arabia, it totaled just 86 units.

Why didn't Electro-Motive sell more F9s? First of all, by the mid-1950s, carbody-style locomotives had fallen out of favor. Streamlined full-carbody locomotives were viewed as

In 1976, Conrail was formed from bankrupt eastern carriers including Penn-Central. The only F7A to receive Conrail blue paint was No. 1648, a former New York Central locomotive. (photo by Doug Eisele)

POSTWAR Fs   **61**

In late 1969 or early 1970, Penn-Central bought 10 former Rio Grande F units, intending to trade them in on new GP38s. However, at least four were pressed into service with PC "noodles" and numbers. At Ohio's Collinwood Yard on February 22, 1972, PC No. 754, ex-Rio Grande No. 5754 offers a bit of color to a fleet of dreary, black locomotives. (photo by James P. Marcus, Doug Eisele collection)

Headlight detail on an F7A. (photo by Brian Solomon)

Milwaukee Road was one of several lines that had F units equipped with nose-mounted "lift-lugs" used for maintenance. (photo by Brian Solomon)

The National Railroad Museum at Green Bay, Wisconsin, has three F units on display. This one, dressed as "Janesville & Southeastern" No. 106, is a former Milwaukee Road F7A, originally numbered 117C. All three F units came from the collection of the late Glen Monhart. They were previously stored at the roundhouse in Janesville, Wisconsin. (photos by Brian Solomon)

Boston & Maine F7A No. 4266 was built in 1950 and was among the last B&M F units retired. B&M's B units lasted longer in regular service than its A units; 4267B remained active until 1979. (photo by Brian Solomon)

Electro-Motive F units operate to virtually every corner of the North American railroad network. Alaska Railroad F7A No. 1506 and F7B No. 1503 lead the company's executive train at Portage, Alaska, on September 4, 1985. (photo by Tom Carver)

Former Alaska Railroad F7A, under private ownership, was leased to the Massachusetts Central Railroad during the mid 1990s. This 26-mile short line used it for both freight service and passenger excursions. (photo by Brian Solomon)

Former Alaska Railroad F7A No. 1508 works a Santa Train on the New England Central at Three Rivers, Massachusetts, in December 1996. This locomotive is equipped with an extra-large plow to cope with Alaska's extreme snowfall. (photo by Brian Solomon)

necessary for road operations in the early days of dieselization, but by the time the F9 debuted, most American railroads preferred road switchers. These were cheaper, more versatile, and easier to maintain in a shop. Also, they didn't require turning when operating singly. In the early days of mass-produced road-diesels, Electro-Motive set developmental trends. It promoted its sleek, sexy, powerful, and reliable Es and Fs. While Alco and Baldwin offered road switchers, Electro-Motive didn't bother with this type since it was basically selling as many locomotives as it could build anyway. In the late

Privately owned former Alaska F7A No. 1508 had a 16-567C engine installed at Palmer, Massachusetts, in December 1993. Although the 16-567C can produce 1,750 hp, the 1508's electrical system was not upgraded, and the unit retained its 1,500-hp rating. (photo by Tom Carver)

1940s, it introduced road switchers to its catalog. First was the odd-looking BL1/BL2, which was essentially an F3A in modified carbody better suited for switching, then later came the GP7, which used most of the same equipment as the F7. The success of this locomotive type is demonstrated by the numbers sold. According to Louis A. Marre's *Diesel Locomotives: The First 50 Years*, Electro-Motive sold 2,610 GP7s during its production run between 1949 and 1954. The GP9 quickly exceeded GP7 totals and became its best-selling locomotive — more than 4,000 were built for the North American market.

Electro-Motive's road switchers could be operated in multiple with F units. Some railroads, notably

In 1997, former Alaska Railroad No. 1508 was moved to the Adirondack Scenic Railroad, based in Utica, New York. The unit was repainted in a livery inspired by the New York Central "Lightning Stripe" scheme used during the 1940s and 1950s. (photo by Brian Solomon)

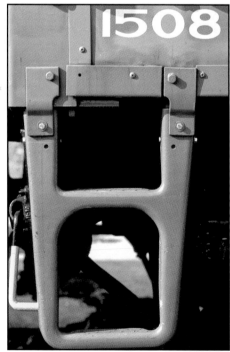

Rear step on Adirondack Scenic Railroad F7A No. 1508. (photo by Brian Solomon)

Norfolk & Western and Illinois Central, skipped the F unit altogether and replaced steam en masse with road switchers (both lines later inherited Fs through mergers, N&W through Wabash, IC through Gulf, Mobile & Ohio).

Another factor affecting locomotive sales in general was simple market saturation. During the mid-1950s, most American railroads completed planned conversions from steam to diesel. The frantic pace of locomotive purchase precipitated by the end of World War II had finally subsided. By the late 1950s, one of the largest markets was providing replacements for World War II-era Fs. By that time, high-horsepower road switchers, both four-axle and six-axle models, were the choice locomotive styles.

The few American railroads that bought F9As were: Louisville & Nashville, Milwaukee Road, Northern Pacific, Rio Grande, Santa Fe, and the Minnesota iron ore hauler Erie Mining. In addition, NdeM bought a few. Northern Pacific had by far the largest roster with 37 units, some geared for freight, others for passenger service. The F9Bs were more widely sold, and in addition to the above-mentioned lines, Clinchfield, Frisco, Great Northern, and Katy also bought them, as did both Canadian National and Canadian Pacific.

All production F7As used this style of number boards and class lamps. (photo by Brian Solomon)

Rear body view of Adirondack Scenic Railroad F7A as seen on the platform at Utica, New York, in July 2004. (photo by Brian Solomon)

This is a detailed view of the sand fill hatch on Adirondack Scenic Railroad F7A No. 1508. (photo by Brian Solomon)

The engineer's side cab interior detail for Adirondack Scenic Railroad F7A No. 1508 shows gauges for traction motor load, airbrake pressure, and warning lamps. (photo by Brian Solomon)

This interior view of Adirondack Scenic Railroad F7A No. 1508 shows the value of the porthole windows in the carbody. This is a 16-567C diesel engine; F7As were built with the earlier 16-567B engine. (photo by Brian Solomon)

The 16-567C produces 250 hp more than the older 16-567B engine and is a more reliable design. It can be easily distinguished by its round hand-hole covers in place of square ones used on older models. (photo by Brian Solomon)

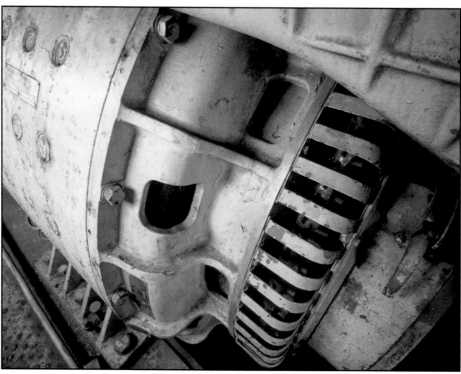

Electro-Motive model F7A uses a D27 generator. (photo by Brian Solomon)

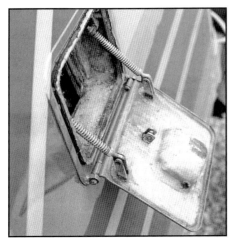

This sand fill hatch on the engineer's side of former Alaska No. 1508 is in the open position. (photo by Brian Solomon)

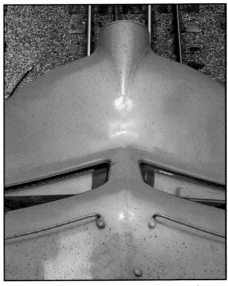

This down-on view of the nose of F7A No. 1508 provides visual clues to how the sheet-metal components are fitted together. (photo by Brian Solomon)

Close-up view of a nose louver on Adirondack Scenic Railroad No. 1508. (photo by Brian Solomon)

Postwar Carbody Changes

For the sake of casual identification, locomotive chronologists have delineated distinctive external features that may be used to distinguish one model from another. In some cases, external modifications are affectations of more important internal changes such as improved filtration or airflow. Other changes were designed to enhance aesthetics or cut unit production costs.

The original *Diesels Spotter's Guide* by Jerry Pinkepank and Louis Marre published in 1966, and a Don Dover's article "All About F's" in the January 1970 *Extra 2200 South*, were among the first to chronicle changes and "phases" in Electro-Motive F units. However, as far the manufacturer was concerned, carbody changes do not necessarily distinguish models from one another. The phases listed here are those essentially delineated by Don Dover and have subsequently become part of railroad literature. They are not manufacturer's designations, nor do they necessarily represent buyers' options on the different models.

The F3 exhibited four different carbody phases over its three-year production run. The earliest, designated phase I, was shared with the F2. These two Fs appear externally identical. The distinguishing features of this carbody style are three evenly spaced side porthole windows. Later F A units have just two windows (B units retained the three-window arrangement through the end of F9 production). Side air intakes are located entirely above the portholes and are covered with a broad wire mesh, sometimes described as "chicken wire," although in reality it is a much heavier mesh than that used for penning poultry. Number boards are of the smaller side-mounted variety used on the FTs and some E units. This body style was used until approximately May 1947.

Phase II F3s introduced the larger, boxy number boards on the nose. They have just two portholes, feature four

Cab interior of former Alaska Railroad No. 1508. The black 26L brake stand on the left is a modern addition to this 1953-built F7A. (photo by Brian Solomon)

This Blomberg truck detail shows the speedometer cable, journal bearing, and brake cylinder. (photo by Brian Solomon)

The nose door on an F unit opens inward. (photo by Brian Solomon)

Electro-Motive F units shared some components with parent General Motors' automobiles. The window handles inside the cab of F7A No. 1508 are essentially the same as those used on GM cars in the early 1950s. (photo by Brian Solomon)

filtered air vents between the portholes, and the wire mesh extends down between the portholes, covering the air vents. This body was in vogue until late 1947. Phase III units use sheet-metal louvers in the carbody to cover lower air vents and wire mesh for the large intakes higher up. The phase III units were the least common. This style was built from late 1947 until the summer of 1948. For the most part, phase I, II, and III F3s featured four raised, shrouded radiator fans, with dynamic brake grids inside the carbody and flush openings behind the cab.

The carbody used by phase IV F3s and phase I F7s is one of the most

---

68   **EMD F-UNIT LOCOMOTIVES**

Later F7s and all F9s use 48-in. dynamic brake fans such as that pictured here on F7A 1508. Some F3s and F7s featured 36-in. dynamic brake fans, the same diameter as the radiator exhaust fans. (photo by Brian Solomon)

Looking from the back of the roof toward the front of No. 1508 provides a good view of the winterization hatch over the rear radiator fan. (photo by Brian Solomon)

common and best-recognized F-unit bodies. It was used on F3s from mid-1948 until the end of production in early 1949, and on F7s until late 1951. It was the first carbody phase to use the lower, non-shrouded radiator fans (diameter 36 in., same as the shrouded fans on earlier Fs), and introduced large stainless-steel side grills over the primary air intake vents. This style of grill is characterized by three distinct rows of six narrow horizontal slits punctuated by vertical cross members situated every two feet. The lower air-intake louvers, between the side porthole windows, were made from carbody sheet metal like those on phase III F3s. A 36-in. dynamic brake fan was also introduced during this production period, and a second headlight located in the nose door was available as an option. Some railroads availed of the second headlight, using the nose mounted headlight as either a white oscillating headlight (often manufactured by Mars, and thus known colloquially as "Mars Lights" regardless of brand), or as a red oscillating warning light.

Phase II F7s were produced between the end of 1951 and the end of F7 production in December 1953. The trademark of this phase was the introduction of a modified air-intake grill made of stamped stainless steel and featuring three rows of vertical slits instead of rows of horizontal slits on the older grills. This style of

On September 19, 1990, former Alaska Railroad FP7s Nos. 1510 and 1512 were at work on the Wyoming Colorado Railroad at Walled, Colorado. Note the extra large winterization hatches. (photo by Brian Jennison)

Southern Railway FP7 No. 6145 leads train No. 6. The FP7 was 4 ft. longer than a normal F7A to make room for additional steam-heating equipment and was designed for long-distance passenger runs. (Doug Riddell collection)

Western Pacific's FP7s were routinely assigned to the railroad's premier luxury passenger train, the California Zephyr. On August 26, 1967, Western Pacific 805D basks in the sun at Middle Harbor Road in Oakland, California. (photo by Roger Puta, Brian Jennison collection)

Western Pacific's nearly new FP7 No. 804-D glimmers beneath the gloom of Oakland Pier's ancient sheds. The California Zephyr connected Chicago and Oakland, California, via the Burlington, Rio Grande, and Western Pacific. (photo by Fred Matthews)

Leading the stainless-steel streamlined Crusader, Reading FP7 No. 905 departs Central Railroad of New Jersey's Jersey City terminal in the early 1960s. The New York City skyline looms across the Hudson. (photo by Richard Jay Solomon)

Reading had just three FP7s, which were often assigned to its famous Jersey City-Philadelphia Crusader. This train is seen on its final lap to Jersey City. (photo by Richard Jay Solomon)

grill was also used on all U.S.-built F9s, FL9s, and some Canadian F9s and FP9s. A few Canadian-built Fs used a third type of stainless-steel grill, stressing vertical openings without horizontal bars. This style was used on FP7s as well as F9 types. Other changes introduced during phase II production include a larger dynamic brake grid fan (48 in. diameter), the use of vertical slits on lower side air-intake louvers (each used two rows of narrow-stamped vertical slits instead of a single row of wider horizontal slits), and the discontinuation of the stylized sheet-metal overhand at the rear of the roof. This sheet metal was strictly styling and had no effect on performance.

The F9 body style is virtually identical to later F7s but is distinguishable by the addition of another side louver between the cab door and first porthole window, and by the use of a flush headlight design. On older F units, the headlight housing features a rounded lip over the glass.

Wisconsin & Southern FP7 No. 71A leads a football special at Madison, Wisconsin, on September 23, 2000. This is one of many former Milwaukee Road Fs that have outlived the Milwaukee itself. (photo by Brian Solomon)

In its early years, Amtrak operated former Southern Pacific FP7s in passenger service. On April 12, 1975, a Pacific Locomotive Association excursion negotiates California's Niles Canyon. (photo by Brian Jennison)

Former STCUM FP7 No. 1305 now works for the Quebec Central short line. It leads a freight at Broughton, Quebec, on January 27, 2004. (photo by Don Marson)

Into the 1990s, Montreal commuter rail operator Societe de Transport de la Communaute Urbaine de Montreal (STCUM) operated its suburban trains with former Canadian Pacific FP7s. In January 1993, No. 1301 approached Dorval, Quebec. (photo by Brian Solomon)

Canadian Pacific FP9 No. 1406, seen at Calgary, Alberta, on August 20, 1971, has an unusual style of radiator intake grills used on some late-era Canadian-built F units. (photo by Bill Dechau, Doug Eisele collection)

VIA Rail No. 40, the eastward Atlantic, at Brownville Junction behind former Canadian Pacific FP7 No. 1414, on August 1979. (photo by Don Marson)

POSTWAR Fs

CP Rail restored a set of F units into its maroon and gray livery for use on excursion and business trains. (top photo) On June 5, 2004, two of the three Fs lead an excursion at Waterdown, Ontario. (bottom photo) These units have a classic look, yet have modern features such as ditchlights. (photos by Pete Ruesch)

LTV Steel's former Erie Mining iron ore railroad in Minnesota's Iron Range was one of the last places to find A-B-B-A sets of F units hard at work in freight service. This railroad was built new in the mid 1950s and was among the last to purchase Fs for freight work. Extremely harsh winter weather made an enclosed carbody more appealing. LTV Steel's F9s work ore trains on September 24, 1994. (photos by Brian Solomon)

Close-up of the distinctive flush headlight style used on F9As, FP9s, and FL9s. Compare this to the more rounded headlight lips on earlier F units. This former Canadian National FP9 operates on the Conway Scenic Railroad. (photo by Brian Solomon)

A pristine Canadian National FP9 and an F B-unit pose in the sun on June 1961. Notice the steam-era bell mounted on the roof. (photo by Richard Jay Solomon)

POSTWAR Fs

CHAPTER THREE

NEW HAVEN FL9

New Haven 2042 and 2041 whisk a Grand Central-bound passenger train past the New York Central coach yards at Mott Haven in The Bronx, New York, circa 1961. (photo by Richard Jay Solomon)

Electro-Motive's longest, last-built, and by far its most unusual F unit was the FL9 model built exclusively for the New York, New Haven & Hartford Railroad (New Haven). The FL9 was a highly specialized machine designed for a specific application. It was a unique design, with specifications carefully tailored for New Haven's unusual service requirements.

By the turn of the 20th century, the New Haven Railroad had pioneered the mainline application of high-voltage alternating current overhead electrification. Beginning in 1905, New Haven began what would become one of the most intensive mainline electrification systems in the United States. Its wires ultimately reached from the New York Metropolitan area east on its multiple-track mainline to its Connecticut namesake, New Haven. Branches to Danbury and New Canaan, Connecticut, as well as some short freight-only lines, were also electrified. New Haven's electric operations included its intensive suburban passenger service, frequent Boston-New York long-distance passenger services, other long-distance passenger trains, and its very heavy freight traffic.

The impetus for electrifying came from New York City's ban on the operation of steam locomotives. This came in response to a horrific crash in the Park Avenue Tunnel that killed 15 people. To reach Grand Central Terminal in New York City, New Haven used trackage rights over New York Central, and following that line's electrification, New Haven's trains drew power from Central's under-running direct-current third rail.

New Haven had more ambitious goals for electrification than New York Central. It hoped to electrify the length of its mainline to Boston, as well as

Lurking in Grand Central Terminal's subterranean passages below Manhattan is FL9 No. 2017 on February 10, 1958. In New York Central's electric territory underground, FL9s draw power from the under-running third-rail seen to the left of the locomotives. (photo by Jim Shaughnessy)

New Haven's dual-mode FL9s allowed the railroad to retire its older straight overhead electric locomotives. In July 1964, three sets of FL9s sit where New Haven once stored its EP2, EP3, and EP4 passenger motors. (photo by Jim Shaughnessy)

its New Haven-to-Springfield, Massachusetts, mainline. These plans were curtailed when company finances stalled during the World War I period. New Haven's fortunes never returned, and while the railroad had some of the densest mainline traffic in the United States, it continued as a weak financial performer.

New Haven remained the end of the wire for almost 90 years, a major ter-

It's July 1963, and a pair of FL9s race through South Norwalk toward New Haven, Connecticut. New Haven's FL9s were largely used on Boston-to-New York trains, eliminating the need for the New Haven engine change. (photo by Richard Jay Solomon)

minal and point where locomotives were exchanged on trains traveling on to Boston or Springfield. Although inconvenient, exchanging locomotives was accomplished relatively quickly in the steam era. Also, since New Haven is the eastern terminus of the railroad's New York suburban services, many trains terminate here. To a lesser extent, the New Haven Railroad also had an engine change at Danbury for trains on the line to Pittsfield, Massachusetts.

After World War II, New Haven's electrification was nearly four decades old and showing the signs of age and exceptionally heavy use. The railroad searched for ways to curtail expenses, and during the decade following the war it faced tough decisions.

New Haven had been an early proponent of dieselization, and on the eve of World War II had invested in a fleet of Alco DL109 road diesels. The road had completely phased out steam operations by the early 1950s. Yet, unlike many American railroads, New Haven had not embraced Electro-Motive, having bought diesels from Alco and Fairbanks-Morse. During the heyday of the Electro-Motive F unit, this quintessentially American model was ignored by New Haven.

By the mid-1950s, modern diesels had clearly demonstrated their superior efficiency as railroad motive power. Low cost of diesel fuel and relative efficiency of diesel-electric power had made additional investment in overhead electrification unnecessary.

By the 1950s, many of New Haven Railroad's old electrics were worn out, and its DL109s and other early diesels were approaching retirement. New Haven believed modern diesel technology could allow it to scale back its overhead

Penn-Central renumbered New Haven's FL9s into the 5000-series number block. Number 5043 was originally New Haven 2043, built in 1960. The later FL9s used 16-567D1 engines instead of 16-567Cs. (photo by George Kowanski)

New Haven Railroad 2008 and 2012 hum through Williamsbridge, New York, on June 30, 1959. These locomotives were part of the first order of FL9s built in 1956 and 1957. An additional 30 were built in 1960. (photo by Richard Jay Solomon)

electrification and replace its early diesel fleet. First it phased out electric freight operations in the mid-1950s (although electric freight service was re-introduced for a few years in the mid-1960s). Yet, its long-distance passenger operations presented an operating conundrum. Long urban tunnels in New York City precluded diesel operation into Grand Central and Penn Station. New Haven considered

NEW HAVEN FL9

a variety of options, including the possibility of rebuilding its old DL109 fleet into dual-mode diesel-electric/electrics. This option was scuttled, in part because such a locomotive would have exceeded weight restrictions on New York Central's Park Avenue viaduct.

Enter the FL9

In 1956, Electro-Motive built a pair of prototype diesel-electric/electric F units, numbers 2000 and 2001, adapted from the passenger-service FP9. The FL9 is essentially a "hybrid" locomotive. It can operate as a normal diesel-electric using the onboard diesel engine to generate current, or as an electric locomotive drawing current from line-side third rail. The locomotive needed to make the relatively complex electrical transition from diesel-electric to electric while rolling along at speed, which was done with additional electrical equipment housed in the FL9's noticeably longer carbody. At 58 ft. 8 inches, it is a full eight feet longer than the F9 and four feet longer than the FP9. Authors Joseph R. Snopek and Robert A. La May explain in *Diesels to Park Avenue* that the "L" in the model designation stands for "lengthened." Other authors have variously suggested that the "L" indicates "electric" or "lightweight."

Some FL9s were painted in Penn-Central black. Here 5038, former New Haven 2038, passes Shell Tower at New Rochelle, New York, on its way to Grand Central. (photo by George Kowanski)

The FL9 uses the unusual B-A1A wheel arrangement. To spread the weight of steam boiler and added electrical equipment and keep locomotive axle weight in compliance with the Park Avenue Viaduct stringent weight restrictions, the FL9's rear truck uses a three-axle Flexicoil truck with an A1A arrangement — whereby only the two outside axles are powered — instead

In the Penn-Central era, many FL9s were painted in this blue and yellow livery. This photo at Oscawanna, New York, in July 1976 was made just a few months after Conrail assumed Penn-Central operations. The second unit is in darker Conrail blue. (photo by George Kowanski)

Amtrak FL9 485 leads train No. 48, the New York-bound Lake Shore Limited, at Albany-Rensselaer, New York, in February 1989. Amtrak rebuilt six of the twelve FL9s it received with the demise of Penn-Central. (photo by Brian Solomon)

This is a close-up view of an Amtrak FL9's three-axle Flexicoil truck. The center axle is not powered and is used to distribute weight more evenly. The third-rail shoe is suspended from the shoe beam. (photo by Jim Shaughnessy)

of the typical two-axle Blomberg truck typically used on Fs.

Following successful experimentation with the prototypes, largely on New York Central's suburban electrified lines, New Haven ordered a fleet of 28 production units, numbered 2002 to 2029. These were delivered to New Haven during 1957. Electro-Motive implemented a few changes on the FL9 design as a result of prototype testing. As common with other F units, the original prototypes had been equipped with the common four-wheel Blomberg truck in front. However, since this truck was unsuitable for supporting third-rail shoes, Electro-Motive replaced it with a four-axle Flexicoil truck that could carry a wooden

At the end of 1982, Conrail exited the commuter rail business, and a new subsidiary of New York's Metropolitan Transportation Authority was formed — the Metro-North Commuter Railroad. In 1989, a pair of Metro-North FL9s zip north of Peekskill heading toward Poughkeepsie, New York. (photo by Brian Solomon)

It was a dark and stormy night under the sodium vapor lights on September 6, 1987, when this photograph of Metro-North Nos. 2003 and 2024 was exposed. MNR No. 2003 was originally New Haven No. 2017, while No. 2024 was New Haven No. 2058. (photo by Tom Carver)

beam supporting a third-rail shoe. Using this arrangement, third-rail shoes were carried on both front and rear trucks, which gave the locomotive better electrical contact. In addition, a small pantograph, located on the roof of the locomotive, was designed to contact overhead direct-current electric pickup in places where it was impractical to locate third rail on the ground, such as long crossovers using double-slip switches in Grand Central.

New Haven designated these first 30 FL9s as Class EDER-5. Each was powered by Electro-Motive's 16-567C engine, rated at 1,750 hp, and used a D32 generator and four D37 traction motors. Overall, the FL9's electrical system is more involved than that in a typical F unit. Among the additional equipment was protective circuitry to prevent damage from a high-voltage short circuit caused by the third rail. The main generator was powered directly from the engine, so an alternative source for on-board electricity was needed when the FL9 was drawing tractive current from the third rail. This was provided by a motor-generator set with an alternator, which supplied direct current to charge engine batteries and to power an auxiliary air compressor, and alternating current for traction motor blowers and other auxiliary equipment. In

New Haven Railroad FL9 No. 2027 was on its fourth owner and went through as many numbers by the time this image of its number board was exposed in Danbury, Connecticut, in 1989. By then, Metro-North No. 2030, and old New Haven No. 2030, had been scrapped. (photo by Brian Solomon)

addition, an electric water heater was used to keep engine-cooling water hot to ease the starting of the diesel on the move. The locomotive came with the distinctive looking and sounding Hancock air whistle, positioned on a bracket between the windshields.

New Haven's FL9s were painted in the distinctive red, white, and black liv-

Like the F9A from which the model was derived, the FL9 used flush headlights. photo (by Brian Solomon)

ery debuted in the McGinnis-era on the railroad's last new straight-electric locomotives, 10 model EP5s, delivered in 1955. This modern livery is a contrast to the more classic paint schemes designed by Electro-Motive. Instead of gently flowing curves and lines that accented the streamlined shape of the carbody, the McGinnis scheme used geometric shapes and straight lines with harsh, clashing colors.

More FL9s

Despite some electrical difficulties, particularly when operating in third-rail territory, New Haven was pleased with the FL9. It enabled New Haven to retire older diesels and electrics, and eliminate engine changes for most through passenger trains. At one time, the railroad hoped to buy as many as 98 FL9s, but it could only obtain financing for an additional 30 locomotives.

The second order was built in 1960 and numbered 2030 to 2059. The carbody-style locomotive had fallen out of favor in the United States, and New Haven FL9 2059 was the very last F unit built by General Motors. It ended more than two decades of production of one of the best-known and most successful diesel-electric locomotives. (Incidentally, the historic final F unit, New Haven No. 2059, later became Metro-North No. 2033, and has been preserved.) The carbody style continued for three more years, as the Electro-Motive E unit remained in production. Although New Haven was the only railroad to purchase FL9s, both New York Central and Pennsylvania had considered operating dual-mode diesels.

The second order of FL9s incorporated a variety of small changes and improvements. Although locomotives carried the same model designation, New Haven classed them EDER-5a to distinguish them from the first 30 units. Electro-Motive had upgraded to its 567-diesel design, designating its latest model 567D1, which delivered 1,800 hp, representing a nominal increase over the 567C. The new engine also incorporated an improved cooling circuit. As a result, no external cooling pipes were on the roof of the locomotive.

Snopek and La May explain that New Haven was on a tighter budget with the second order. Unlike the first 30, the second batch did not feature head-end multiple unit connections, roof-mounted pantographs, or dynamic brakes.

Another minor external change was the introduction of a narrow foothold to reach the nose from the cab. This is not a specific spotting feature, since earlier units were later retrofitted with footholds.

FL9s at Work

New Haven routinely assigned FL9s to passenger services between Boston and Grand Central, and occasionally to Penn Station. They also regularly worked trains to Springfield. Later in the New Haven era, FL9s worked trains up along the Housatonic River with trains for Pittsfield, Massachusetts, via Danbury, Connecticut. They also worked some trains to Cape Cod. Although an uncommon task, FL9s also occasionally worked freight.

In 1969, the financially destitute New Haven was absorbed by the recently created Penn-Central (a gigantic line formed through the combination of New York Central and Pennsylvania). PC control greatly altered FL9 assignments. Many FL9s were reassigned to former New York Central Grand Central suburban services, establishing a pattern that remained for decades despite tumultuous changes in Northeastern railroading.

This derelict FL9 stored at Rome, New York, in July 2004 has several missing side panels, which gives a good view of the interior trussing and location of electrical cabinets at the back of the FL9. (photo by Brian Solomon)

Working in "push" mode, Metro-North FL9-AC No. 2042 shoves a Grand Central-bound train along the Hudson Line near Manitou, New York, on June 27, 1997. This was one of seven FL9-ACs rebuilt by ABB-Traction for Metro-North. It was originally New Haven Railroad No. 2022. Note the lack of side porthole windows. (photo by Brian Solomon)

Penn-Central painted some FL9s in its minimalist glossy black. Others were dressed in dark blue with bright yellow noses. A few rattled around in faded and patched former New Haven dress for many years before receiving new paint. Some never saw fresh paint and were cut up in New Haven colors.

In 1971, Amtrak assumed long-distance passenger services from Penn-Central and 16 other railroads around the United States. Amtrak did not initially own or operate any FL9s, and in early years assigned E units to most of its long-distance trains in non-electrified territory. Penn-Central's financial collapse in the early 1970s ultimately resulted in many of the company's passenger-rail assets being transferred to Amtrak, while the company's freight and suburban passenger operations were conveyed to Conrail in 1976. Among Amtrak's acquisitions were 12 FL9s, half rebuilt for service with the remainder scrapped or used for parts. Amtrak assigned its FL9s to work on the Hudson Line in Empire Corridor service. Initially, they were used to replace worn-out, antique former New York Central third-rail electrics between Croton-Harmon and Grand Central. In the mid-1970s, Amtrak moved the engine change from Harmon, up the river to Albany-Rensselaer. Amtrak's six FL9s spent more than two decades in Hudson Line service, whisking through passenger trains along the shores of the river. Amtrak invested in having its six FL9s rebuilt, which extended their usefulness. Details of the rebuilding are covered below.

Conrail assumed Penn-Central's suburban operations on April 1, 1976, along with the freight operations of PC and a half-dozen other lines. Conrail exited the passenger business at the end of 1982, and former New York Central and New Haven services to Grand Central were conveyed to Metro-North. With the transfer operation, Metro-North inherited many former New Haven FL9s. Among these were four owned by the Connecticut Department of Transportation, used on through branch-line services on former New Haven Railroad branch lines in Connecticut. These units were rebuilt by Chrome-Crankshaft and repainted in the traditional McGinnis livery in 1985. By the mid-1990s, ten FL9s had been restored to this livery. In theory, these locomotives were assigned to Connecti-

This panned view of Metro-North FL9-AC No. 2041 gives a good view of the model's tapered roof profile and modified side panels, both a result of extensive rebuilding by ABB-Traction. The FL9-ACs use 12-710G engines and a modern three-phase AC traction system. (photo by Brian Solomon)

EMD F-UNIT LOCOMOTIVES

cut services, but in practice they often worked elsewhere on Metro-North.

Electrification of the Harlem Line to Brewster in 1983-1984 freed up many FL9s for other services. For several years in the mid-1980s, Metro-North was short electric-multiple units on the New Haven run, and pairs of FL9s hauling steam-heated passenger cars (some borrowed from other suburban railways) were assigned to work rush-hour GCT-New Haven Line services.

Metro-North, like Amtrak, extended the life of its FL9s, contracting to have them rebuilt and modernized. Although many of Metro-North's FL9s have been sold or retired, in 2004 a few remained active, making them among the last F units to earn revenue for purposes other than tourism or history.

Rebuilt FL9s

After decades of demanding service and less-than-ideal maintenance conditions, the remaining FL9s were in desperate condition. By the late 1970s, these machines were well worn, and many looked much older than they were. Had these FL9s simply been ordinary locomotives and easily replaced by existing production models, in all likelihood most would have been cut up. However, their special electrical and weight characteristics made them too valuable to scrap — especially at a time when passenger railroading was in a precarious financial position.

Between 1978 and 1980, Amtrak contracted Boise, Idaho-based Morrison-Knudsen to overhaul its FL9 fleet, while preparing them for head-end power (HEP — a modern electrical system for train heat and lighting). For the most part, original equipment was retained and rehabilitated. Some new components were used, and a number of external changes were implemented. Snopek and La May detail the changes to these six locomotives. On the electrical end, the traction motors were upgraded to D77Bs. Inside the carbody, provisions were made for installation of head-end

Connecticut Department of Transportation FL9 No. 2019 leads a Poughkeepsie, New York-bound train near Breakneck Ridge on November 20, 1992. This was one of four FL9s rebuilt by Chrome Crankshaft in 1985. (photo by Brian Solomon)

As a cold November day dawns, CDOT FL9 No. 2023 waits with a Bombardier-built push-pull train at the old New Haven Railroad station in Waterbury, Connecticut. This locomotive was built as New Haven No. 2057, and was later Penn-Central No. 2057. (photo by Brian Solomon)

power. Auxiliary diesels and HEP generators were eventually installed in all of these locomotives. External changes included a five-chime horn replacing the traditional Hancock air whistle, a sealed twin-beam headlight, a pair of strobe lights located on the cab roof, and a red warning light installed on the nose. The locomotive bodies were painted in Amtrak's "platinum mist" with a wrap-around red, white, and blue stripe located about midway up the side of the

locomotive. Equipment below the body, including trucks, plus the cab, top of nose, and roof, were all dressed in glossy black paint, much in the same style of Amtrak's more common F40PH of the same period. The locomotives were numbered 485 to 489, plus 491, which was later renumbered to 484 following subsequent rebuilding in 1990.

Between 1990 and 1993, Amtrak overhauled its FL9s again, this time performing the work in-house at its Beech

One of the more successful FL9 rebuilding jobs was by Morrison-Knudsen on the FL9Ms in the early 1990s. A pair of CDOT FL9Ms lead a Wassaic-bound train through Towners, New York, on January 18, 2003. The former New Haven Maybrook Line crosses over the former New York Central Harlem line here. (photo by Patrick Yough.)

In January 2001, FL9Ms painted for New Haven and New York Central work a Grand Central-bound Metro-North train at Croton-North, New York. (photo by Brian Solomon)

Grove, Indiana, shops. While the majority of work was performed to keep the locomotives in good running order, there were again some notable changes with equipment. The traditional engineer's control stand was replaced with "Dash-2" style controls, complete with updated airbrake controls, going from the 24RL schedule to the modern 26L brake schedule. The 16-567C engines were replaced with the 16-645E model. The radiator hatches were adapted for more modern radiators.

Amtrak finally retired its FL9s a few years after new dual-mode General Electric GENESIS locomotives were bought for Empire Corridor services. The two types coexisted for several years. As mentioned, Metro-North also benefited from rebuilt FL9s. The first were seven overhauled by General Electric in 1979.

More significant were four FL9s owned by the Connecticut Department of Transportation, which were sent to Chrome Crankshaft (operating at the former Rock Island shops at Silvis, Illinois) for what Snopek and La May describe as an "in kind" rebuilding during 1984 and 1985. Internally, these locomotives were substantially modernized, and more importantly they were configured to operate in what has become the standard North-American suburban railroad push-pull arrangement. A locomotive is positioned at one end of a set of commuter passenger cars; at the other end is a cab control car from which the engineer can control the train. This arrangement alleviates the need for the locomotive to run around the cars at terminals, greatly reducing turnaround time, simplifying operations, and eliminating the need for turning facilities. One potential downside to push-pull trains is that specialized locomotives are required for the service. In the case of CDOT and Metro-North, this consideration doesn't matter since specialized locomotives are needed anyway.

Although New York Central considered the FL9, it never bought any. In 1999, Metro-North painted two FL9s, 2012 and 2013, in full New York Central lightning stripes, more than 30 years after the railroad was folded in to Penn-Central. The 2013 is shown fresh from the paint shop at North White Plains. (photo by Patrick Yough.)

Amtrak assigned its six FL9s to long-distance trains operating on the Hudson Line to New York City. November 20, 1992, was frosty and cold when Amtrak 488 was photographed near Peekskill, New York. (photo by Brian Solomon)

A detailed view of the third rail shoe mechanism on the rear truck Connecticut Department of Transportation FL9M No. 2014. (photo by Patrick Yough)

Broad-side view of Connecticut Department of Transportation FL9M No. 2026 at Metro-North's Croton Harmon, New York, shops. An easy external identifier for the FL9M is the colored indicator lights on the sides beyond the cab door. (photo by Patrick Yough)

As listed by Snopek and La May, changes to the CDOT FL9s included head-end power using an auxiliary diesel and alternator to generate power; solid-state electronics installed to replace many traditional relays and electro-pneumatic circuit breakers; an electric block heater replacing the traditional engine coolant heater; and a pressurized engine room to keep out particulates. The locomotives were set up to run as push-pull units with new Bombardier train sets using cab-control cars.

In the 1990s, six FL9s were rebuilt as FL9Ms by Morrison-Knudsen, all repainted in the New Haven livery. For this project, MK hired one of the original Electro-Motive engineers who, using modern solutions, corrected shortfalls with the FL9's electrical system that had limited their usefulness in electric mode. A decade later, FL9Ms remain in revenue service on Metro-North and are now among the last F units regularly working in North America.

The most radical FL9 rebuilding, although not the most successful, was the FL9-ACs project. Metro-North's interest in this endeavor began in 1986, and in 1987, the American subsidiary of ASEA-Brown Boveri, ABB-Traction, was contracted to completely rebuild ten units using state-of-the-art three-phase traction technology. During this rebuilding, very little of the original F-unit technology and machinery was retained. Despite their appearance, Santa Fe's CF7 conversions (discussed in Chapter 4) retained more original F-unit parts than the FL9-ACs.

In place of the traditional 16-567 engine, the modern Electro-Motive 12-cylinder 710G was used. More significant was the use of a solid-state controlled three-phase alternating-current traction system in place of the traditional direct-current system. Each locomotive had three inverters, which use solid-state gate-turn-off switches (GTOs) to convert direct current in modulated alternating current — two for traction and one to supply 480 volts AC at 60Hz for HEP. This arrangement eliminated the need for an auxiliary engine and alternator.

In addition to extensive internal changes, ABB-T rebuilt the carbody structure to better suit the new equipment. Among the external changes were replaced side panels and elimination of the side porthole windows, a side equipment door placed in the rear third of the carbody, radiator and exhaust fan arrangement altered and relocated toward the back of the locomotive, and adjustment to the roof profile. A modern vertical-sealed beam headlight replaced the traditional F9-era headlight. Seven FL9-ACs, numbered 2040 to 2046, were rebuilt for Metro-North. The remaining three were built for Long Island Rail Road, numbers 300 to 302. Interestingly, LIRR 301 was rebuilt from pioneering FL9 prototype New Haven No. 2000.

The nose of FL9M No. 2011 basks in the late sun at South Norwalk, Connecticut, on June 2, 2002. (photo by Patrick Yough)

CHAPTER FOUR

RECYCLED F UNITS

The Pioneer Valley Railroad based in Westfield, Massachusetts, is one of more than 50 operators of former Santa Fe CF7s in North America. On July 9, 2004, PVRR No. 2647 poses at Westfield. (photo by Brian Solomon)

The typical diesel locomotive life spans 15 to 20 years. When a locomotive works in hard service, it reaches a point of needing to be rebuilt, retired, or converted for other uses. The FTs were the first F units and thus also the first to be retired in large numbers. In the mid-1950s, several lines traded their FTs back to General Motors for credits on new locomotives, often GP9s. By the 1960s, many of the later F units had nearly reached the end of the usefulness. Electro-Motive looked to rejuvenate their locomotive sales, which had sagged dramatically following the successful dieselization of American railroads, by encouraging railroads to trade in F units on newer, more powerful, and more reliable road switchers. Railroads traded Fs for GP20s, and then later GP30s, GP35s, and six-motor models. In 1965, Electro-Motive introduced a whole new line of locomotives powered by its new 645 diesel, stepping up the purge of F-unit fleets. The SD40 and GP40 models were rated at 3,000 hp, thus each locomotive could effectively replace two 1,500-hp F3s or F7s. By the early 1980s, most F units had been traded in or scrapped. However, some railroads chose to recycle old Fs, converting them for other purposes.

Santa Fe remanufactured many of its old F7s into road switchers, designated CF7s. Southern Pacific and Burlington Northern were among railroads that converted F units into generator cars to power rotary snowplows. A few railroads rebuilt Fs into road-slugs (railroad motive power adjuncts equipped with traction motors and ballast, but no primemover and operated in conjunction with a locomotive to provide extra tractive effort). Some Fs were rebuilt for modern suburban commuter service. In

In the early morning of September 19, 1991, Pioneer Valley CF7 No. 2597 rests at the yard in Westfield, Massachusetts. Pioneer Valley operates former Conrail branches in western Massachusetts. (photo by Brian Solomon)

Massachusetts Central operates the former Boston & Albany Ware River Branch between Palmer and South Barre, Massachusetts. In 1984, it acquired former Santa Fe CF7 No. 2443, originally Santa Fe F7A No. 306L, seen here at South Barre on October 16, 1993. (photo by Brian Solomon)

Amtrak's 25 CF7s were numbered 575 to 599 and acquired in trade from Santa Fe in 1984. No. 579 was assigned to a work train at New London, Connecticut, in November 1992. (photo by Brian Solomon)

some cases they were fitted with head-end power assemblies and continued to work as locomotives. In other cases they were stripped of traction motors and related components and used as head-end cab-control cars; these have a dual function with primemovers used to provide head-end power for passenger cars and cabs for service in "push" mode on bi-directional equipment with a locomotive at the far end. Other reused Fs have been made into cars carrying radio-controlled equipment for remote-control helpers, and as steam generator cars for passenger train heating.

Santa Fe CF7

Santa Fe operated one of the largest F fleets, having purchased large quantities of FT, F3, F7, and F9s, using them in all varieties of mainline services. Santa Fe's locomotive maintenance was considered among the best in the industry, and even after 15 to 20 years of hard service, its F primary components were in comparatively good shape. As part of a larger rebuild program to extend the life of its older Electro-Motive diesels, in 1970 Santa embarked one plan to convert some F units from carbody locomotives to road switchers. In so doing it hoped to get another 10 to 12 years from them, while making the locomotives more useful for switching. High-horsepower models

This frontal view of Amtrak No. 579 shows the later cab style. Although it was boxy and less appealing to look at, the roomier cab was preferred by crews. This locomotive had been Santa Fe No. 2437, converted from F7A No. 237L in September 1977. (photo by Brian Solomon)

EMD F-UNIT LOCOMOTIVES

Former Mid-South CF7 No. 7003, operated by Louisiana & Delta, rumbles a short freight down the street at New Iberia, Louisiana. Former Santa Fe No. 2621 was rebuilt from F7A No. 227L in May 1972. Its contour cab was later replaced with the larger boxy cab as pictured here. (photo by Brian Solomon)

had bumped F units from premier road assignments. Since F3s and F7s were rated at a modest 1,500 hp, by 1970 they were better suited for switching chores and secondary mainline work than priority assignments. During this same period, Santa Fe also converted some F3 and F7 B units into radio-controlled slave powercars for use on Locotrol remote-controlled helper sets.

According to a report by Joe McMillan in the November-December 1969 *Extra 2200 South*, in late 1969 Santa Fe started work on prototype road switcher using F7A 262C. It was intended to resemble an Electro-Motive GP7/9 and was tentatively designated "GP8." Santa Fe's Cleburne, Texas, locomotive shop did the work, and the prototype was released in February 1970. It was now numbered

Louisiana & Delta maintains its CF7 fleet at its New Iberia, Louisiana, shops. L&D No. 1503 basks in rich winter sun on January 22, 1996. L&D's CF7s are numbered to reflect their 1,500 horsepower rating. (photo by Brian Solomon)

Close-up of the top headlight on Pioneer Valley CF7 No. 2558. The lens shade below the lamp is designed to reduce glare in the locomotive cab. (photo by Brian Solomon)

Air-hose connections on the front left side of Pioneer Valley No. 2558 are needed for airbrake connections in multiple-unit operation. The hoses from left to right are main reservoir, actuating, and application and release. The fourth connection, not used, is for sanding. (photo by Brian Solomon)

2649, and following months of testing Santa Fe deemed it successful. For the next seven years, Cleburne shops converted the railroad's F units into CF7s — the new model designation used "C" for "converted." Each rebuilt locomotive was numbered in reverse sequence from prototype 2649 on down. By the time the program concluded in March 1978, Cleburne had turned out 233 CF7s, the last carrying road number 2417.

Santa Fe's CF7 incorporated most of the F unit's primary mechanical and electric components, but it required a new frame and support construction. Among components salvaged from the F unit were the Blomberg trucks, fuel tanks, battery boxes, center sill, main generator, auxiliary generator, load regulator, equipment racks for oil cooler, oil filter assembly, electrical panels, control stand, throttle, and airbrake controls. On some units, a portion of the carbody and cab windows was also retained, giving a hint of the locomotive's F-unit heritage.

A road switcher uses an entirely different structure from a carbody locomotive. To replace the truss car-

Close-up of Pioneer Valley No. 2558's coupler and anticlimber. The short anticlimber sections were cut from the original F-unit body and are used to prevent telescoping in the event of serious collision. (photo by Brian Solomon)

Illinois RailNet uses CF7 No. 3 to work the former Burlington Northern branch north of Flagg Center, Illinois. This locomotive was built as Santa Fe F3A No. 31C and converted to CF7 No. 2636 in April 1973. (photo by Brian Solomon)

body, Santa Fe fabricated a platform frame and built a sheet-metal hood, much like that used on standard Electro-Motive road switchers, to cover primary components. The new configuration required significant relocation of airflow and cooling systems. On postwar F units, radiators and radiator fans are located directly above the primemover. This arrangement is impractical on a CF7, so the radiators and radiator fans were relocated to far ends of the long hood, essentially mimicking the arrangement used by Electro-Motive road switchers. This also required fabrication of all new air-intake vents. Pilots were fabricated, although they incorporated short sections of the F-unit anticlimber placed on either side of the coupler-pockets.

With minor variations between each of the CF7s, it is safe to say that no two locomotives are identical. Furthermore, during the course of production Santa Fe introduced several significant design changes. *Extra 2200 South* issue 65 indicates that only the first eleven CF7s (2649 to 2639) used F-unit side cab windows. From 2638 on down, fabricated side cab windows were used. A more significant cab change came with the abandonment of the carbody cab in later CF7 rebuilds. According to *Diesel Spotters Guide Update* by Jerry A. Pinkepank and Louis A. Marre, in 1974, beginning with CF7 2470, a home-built boxy cab was used in place of the older carbody cab with F-unit contour. *Extra 2200 South* refers to this as the "Topeka" cab. The more common CF7 was the older style, originally representing 179 units, while the Topeka cab CF7s accounted for just 54 units. However, Topeka cabs

This broadside view is of Illinois RailNet No. 3, north of Flagg Center, Illinois, in June 2004. Clues to the heritage of this locomotive are the Blomberg trucks, fuel tanks, and battery box below the frame. (photo by Brian Solomon)

Blomberg truck detail on Illinois RailNet CF7 No. 3 clearly shows the Hyatt journal and airbrake cylinder. (photo by Brian Solomon)

Rear three-quarter-view of Pioneer Valley No. 2647 at Westfield shows the fabricated sheet-metal hood constructed to cover the locomotive's innards in place of the F-unit carbody. (photo by Brian Solomon)

Extra 2200 South, issue 65), a four-stack exhaust was introduced.

The majority of CF7s were rebuilt from F7As; however, the railroad also converted 20 F3As (CF7s as built: 2646, 2642, 2636, 2630 to 2622, 2619 to 2613, and 2609), and 15 F9As (CF7s as built: 2520, 2519, 2504, 2485, 2482, 2480, 2479, 2474, 2471, 2460, 2459, 2441, 2436, 2434, and 2432). Despite the different F-unit models used, all the converted locomotives were designated CF7s, and not "CF3s" or "CF9s." They were rated at 1,500 hp, and used the common 62:15 gear ratio.

During CF7 rebuilding, Santa Fe introduced a variation of its freight paint scheme. The older blue with yellow fronts and end and yellow lettering was replaced by a blue and yellow variation on the Warbonnet, sometime described as the "yellow bonnet." According to *Extra 2200 South* issue 65, the first CF7 in the newer scheme was 2616, built in June 1972.

Between 1984 and 1988, Santa Fe disposed of its CF7 fleet. While some were sold for scrap, many were conveyed

Detail of Hyatt roller bearing on a Blomberg truck. (photo by Brian Solomon)

were more popular with crews, and later Santa Fe retrofitted some older CF7s with them. David Swirk of the Pioneer Valley Railroad described several primary advantages of the Topeka flat roof cab style used on his railroad (which has both carbody cab and Topeka cab units). They have bigger windows, more room inside the cab, better headroom, a more modern control stand with 26L brake schedule, and are noticeably quieter. Among other variations was the use of an open frame versus box frame. Also, the earliest units featured a twin-stack exhaust. Later, beginning with 2613 (according to

Converting the F units to a road switcher format fundamentally altered the structure of the locomotive. To support primary components, Santa Fe fabricated new frames for the CF7s. Below the frame are components, such as the fuel tank, recycled from the original F unit. (photo by Brian Solomon)

to other railroads. The CF7 was a popular machine for short lines looking for inexpensive secondhand Electro-Motive-powered locomotives in the 1,500-hp range. The CF7 is ideal for switching and yard work on lines with light track. More than 50 different companies have used former CF7s. One of the largest fleets was that operated by Amtrak, which in 1984 traded 18 surplus SDP40F road units to Santa Fe for 25 CF7s and 18 SSB1200s (rebuilt switchers). This arrangement suited both railroads. The SDP40F was Amtrak's original EMD-built long-distance passenger diesel, but it had fallen into disfavor in the mid 1970s following a series of highly publicized derailments. Ultimately, Amtrak acquired a fleet of four-motor F40PHs to work its long-distance trains. Santa Fe viewed the cowl-type SDP40F favorably, and integrated the former Amtrak units into its freight pool. Amtrak desperately needed

CF7 radiator intake vents were patterned after those used on early Electro-Motive GPs. (photo by Brian Solomon)

RECYCLED F UNITS

Front steps on the engineer's side of Pioneer Valley No. 2647; among the advantages of road switchers are steps and grab irons that ease switching tasks. The cab is also better suited for viewing switching and operating in reverse. (photo by Brian Solomon)

CF7 handrails are made from 1-1/4-inch pipe, while the grab irons tend to use 3/4-inch pipe. It is necessary to provide grab irons to access the nose and roof. (photo by Brian Solomon)

reliable switchers to replace clapped-out former Penn-Central Alco-built RS3s and Electro-Motive SW1s. Santa Fe's CF7s and SSB1200s fit the bill. Initially, Amtrak operated the units in revised Santa Fe yellow bonnet paint and its own lettering. Later some Amtrak CF7s were dressed in platinum mist. They were assigned to work coach yards and train service on the North East Corridor and elsewhere. Only on very rare occasions have they been used to haul passenger trains.

Air intake louvers on the cab of Pioneer Valley No. 2647. (photo by Brian Solomon)

EMD F-UNIT LOCOMOTIVES

Snowplow Power Packs

Southern Pacific operated some of the most difficult mainlines in North America. Its routes over California's Donner Pass and the Oregon Cascades via Pengra Pass are especially difficult because of extraordinarily harsh winter weather. Storms routinely blanket these mountain ranges with heavy, wet snow. Traditionally, Donner was famous for more than 30 miles of wooden snow sheds, which kept the tracks clear at the highest elevations and areas with heaviest snowfall. The Cascade crossing, completed in the 1920s, was built with long sections of concrete snow sheds.

Southern Pacific operated one of the most famous fleets of Leslie Rotary snowplows, which used an ingenious arrangement of adjustable frontal fan blades to scoop snow and throw it far from the tracks in a continuous unbroken motion. This type of plow is especially useful for clearing deep, heavy snow. Older styles of plows, which pushed snow out of the way, had a tendency to bog down in these conditions. They were vulnerable in rock cuts and in very deep snow that drifted or slid onto tracks. A rotary plow churns away unhindered, clearing drifts and snow slides ten feet deep or more.

Traditionally, rotary plows were powered with on-board steam engines, which turned rotary fan blades. A locomotive is required to push the plow from behind. In the 1950s, when Southern Pacific was phasing out steam operations, it recognized that the diesel-electric arrangement would be better suited for its aging fleet of rotaries. In 1957, SP began converting the best of its steam-era snowplows to electric operation. It replaced reciprocating steam engines with electrical control equipment and used four standard Electro-Motive traction motors to power the rotary fan. Initially it assigned F-unit Bs to snow service, disconnecting traction motors and using

Except for the New Haven FL9s, all F units and CF7s rode on the two-axle Blomberg truck. This close-up view shows the arrangement of the hanger, which carries the swing bolster. The elliptical springs provide suspension for the locomotive. (photo by Brian Solomon)

Rooftop view of Pioneer Valley No. 2558 shows the revised arrangement of radiator fans and engine exhaust stacks used on the earlier CF7s. Compare this with the similar rooftop view of Adirondack F7A 1508. (photo by Brian Solomon)

RECYCLED F UNITS

Looking down on the front fireman's side of Pioneer Valley No. 2558. (photo by Brian Solomon)

The horn on Pioneer Valley No. 2558 is mounted on the cab roof. (photo by Brian Solomon)

Second-hand CF7s provide short lines with flexible and relatively low-cost motive power. With 1,500 hp, this CF7 provides Pioneer Valley with sufficient power for most of its daily switching needs. (photo by Brian Solomon)

Electro-Motive used F9 carbody No. 462 for experimental testing. In the mid-1960s it housed a prototype 645-series engine and the AC/DC electrical system that debuted on 645-diesels in 1965. It was photographed in demonstrator paint at Joliet, Illinois, on November 29, 1974. (photo by James P. Marcus, courtesy of Doug Eisele.)

main generator output to power the plow. Two or more diesels were assigned to shove the rotary plow/B-unit combination. At the end of snow season, SP would restore the B units to road service.

In Southern Pacific Dieselization, John Garmany explains that F units were assigned to snowplow service because the cabs in the carbody units offered operating crews greater protection. Later it was found that six-motor road-switchers were better suited to pushing snowplows as a result of their higher tractive effort.

In 1970, when the F-unit era of Southern Pacific was drawing to a close, the railroad selected nine F7Bs to be rebuilt as snowplow-generating units. According to Garmany, the units were renumbered and permanently reassigned to maintenance-of-way service and thus reclassified as "non-revenue equipment" for tax assessment. The former F7Bs' new numbers corresponded to the plow to which they were assigned. Rebuilding included the overhaul of the 567 engines and electrical alterations, including the removal of high-voltage electrical cabinets. Traction motors and gearing were also removed. The former F7Bs were painted uniform black. Although these were no longer locomotives, they retained many external features that were trademark characteristics of the F unit, including the truss carbody and paneling, stainless steel air intakes, porthole windows, and Blomberg trucks.

Snow service on Donner Pass is among the most difficult work on any American railroad. In wet years, when Pacific storms pound the Sierra Nevada day after day, annual snowfall has reached 800 inches. Normally, the railroad is kept clear using flangers and snow-service Jordan spreaders powered by diesel locomotives. However, when snow becomes too deep, the rotaries are called upon to clear the line and blow snow far from the tracks. The old Leslie plows are still one of the best tools for moving very heavy snow. The re-equipped electrically operated plows maintain constant torque on fan blades, which is better for moving heavy, wet snow than the older steam-powered plows. A steam-powered rotary clearing deep snow is without question one of the most dramatic scenes in railroading, and diesel-electric rotary is second best.

Southern Pacific Leslie Rotary snowplow M207 is powered electrically by former SP F7B 8207, which was originally SP 8299. The B unit is used to generate electricity to power the plow blades, while the plow/B-unit combination is propelled by SP six-motor road units. (photo by Brian Solomon)

On the morning of February 23, 1993, SP Rotary M207 moves tons of snow as it plows westward at Truckee, California. Former F7B 8207 is at maximum throttle to keep the plow from bogging down in heavy snow. Traditionally, the rotary plows were steam powered. (photo by Brian Solomon)

Photographs of the plows blowing an arc of snow cannot convey the thundering noise and great power of these machines. Yet a diesel-electric rotary is not invincible. Ice build-up on the rails or hard-packed snow can still derail one of these monsters. This is one reason why Southern Pacific routinely assigned pairs of rotaries in double-ended sets. Each rotary would have its own converted B-unit generator car, and spliced between the two plow sets would be two to four road locomotives. In later years these were usually six-motor Electro-Motive SD40T-2 or SD45T-2s.

Although in modern times rotary plows have seen very infrequent service — a half-dozen years have passed between assignments — in the end, the rotary plow generating cars outlasted the SP, which merged with Union Pacific in 1997.

Burlington Northern also converted some of its old F units as plow-generating cars, sometimes called "rotary plow powerplants." BN's large fleets of Fs, inherited from its predecessor companies, had worked in revenue freight service into the early 1980s, a decade after SP had retired its freight Fs. BN selected a few for rebuilding as plow generator cars by its shops at Livingston, Montana. Among other modifications, traction motors were removed, electri-

Southern Pacific's snowplow power units must endure exceptionally demanding conditions. The radiator intake on former F7B 8207 is encrusted in ice after three days of freezing conditions on Donner Pass in February 1993. (photo by Brian Solomon)

Burlington Northern converted former Northern Pacific F units into rotary plow power cars during the 1970s. Burlington Northern Santa Fe, former BN power car No. 972577 was originally NP No. 7013C but worked as BN No. 777 in its last years as a locomotive. (photo by Brian Solomon)

cal equipment was arranged to power plow equipment, porthole windows were blocked up, various boxes and hatches were added to the roof, and the units were renumbered and repainted in a plain rust-colored "mineral red" for maintenance-of-way service. Burlington Northern converted both A- and B-units to plow service.

BNSF No. 972577 has had several external modifications since its days as an F B-unit locomotive. Notice the housing on the roof above the engine exhaust, and the blocked-up side porthole windows. (photo by Brian Solomon)

In 1989 and 1990, Burlington Northern converted two snowplow power packs for service as locomotives for its executive passenger train. In June 1996, executive Fs F9A-2 BN-1 and F9B-2 BN-2 were on public display at Galesburg, Illinois. (photo by Brian Solomon)

RECYCLED F UNITS 101

BN's Burlington, Iowa, shop did work on the old Northern Pacific F9s to convert them to modern standards. Externally, there were a few minor modifications, but more substantial changes were made inside the locomotives. (photo by Brian Solomon)

In 1990, almost a decade after conversion to snowplow service, two former F units were re-converted as locomotives for BN's Executive Passenger train. According to the 1980-1991 *Burlington Northern Annual* by Robert C. Del Grosso, BN picked former Northern Pacific F9A 6700A and F9B 7002C, built by Electro-Motive in 1954, and converted to snowplow service by Livingston in the early 1980s, for a complete overhaul by its West Burlington (Iowa) shops. Rebuilding involved numerous external and internal changes. Carbodies were extended at one end, portholes restored, and snowplow service boxes removed. In place of traditional 567 engines and electrical equipment, BN installed new Electro-Motive 16-645E engines, AR10-D14 main generators, D77B traction motors, and "Dash-2" solid-state electrical control equipment. When completed, these locomotives had more in common with contemporary Electro-Motive GP38-2 road switchers and other modern GM diesels than with 1950s-era F units. BN designated its new Executive Fs as "F9-2s" to reflect the electrical upgrade and 645 engine. Traditional-looking Electro-Motive F9 plates were affixed to the F9A. Numbered BN-1, the units were dressed in a classy-looking new livery using dark forest green and cream for the main body color, with fine red stripes and gold lettering, and a brilliant chrome-plated "BN" logo affixed to the nose of the F9A. These locomotives were regularly used on BN's corporate passenger train, but also worked in freight service, sometimes leading the railroad's hot transcontinental intermodal trains.

Burlington Northern's BN-1 was originally Northern Pacific F9A No. 6700A, later BN No. 9800, then BN No. 766, before being converted to snowplow power pack BN No. 972567. (photo by Brian Solomon)

EMD F-UNIT LOCOMOTIVES

Road Slugs

Some F units were stripped of their diesel engines and related equipment and converted into road slugs to be operated in conjunction with other locomotives. A slug is not a locomotive, but works with a locomotive (or locomotives) to provide additional powered axles laden with ballast to provide increased tractive effort without the expense of maintaining additional engines. Typically, a road slug was spliced between two locomotives. Its electrical equipment was operated in multiple units with the diesel locomotives. Electro-Motive F B units were well suited for slug conversion since they were cab-less to begin with.

These old Northern Pacific F9s are back on old home rails; on July 5, 1994, BN-1 and BN-2 lead a westward double-stack container train on the Montana Rail Link near Sand Point, Idaho. This was the former NP main line. (photo by Brian Solomon)

The Electro-Motive builders' plate for BN-1, former Northern Pacific No. 6700A, is dated February 1954. (photo by Brian Solomon)

Massachusetts Bay Transportation Authority FP10 No. 1105 is at Somerville, Massachusetts, north of Boston's North Station in August 1989. Note changes to the roof profile including the raised box around. This area was once occupied by the dynamic brake cooling fan and the muffled exhaust pipes at the back for the HEP generator. (photo by Brian Solomon)

Unlike traditional F units, BN-1 and BN-2 were powered by 16-645 diesel engines rated at 2,000 hp and used modern electrical systems with AR10-D14 main generators. (photo by Brian Solomon)

RECYCLED F UNITS

MBTA 1100, originally GM&O F3A No. 805A, shoves a set of de-powered Budd rail diesel cars at Ayre, Massachusetts, on Aug 14, 1987. This locomotive was later sold to the Cape Cod Railroad. (photo by Brian Solomon)

In March 1989, MBTA No. 1109 rests between runs at the Boston & Maine yard at Fitchburg, Massachusetts. Originally GM&O F3A No. 807B, this locomotive later became Metro-North No. 410. (photo by Brian Solomon)

At 7:58 a.m. on June 15, 1987, MBTA FP10 No. 1153 shoves a Reading, Massachusetts-bound passenger train pauses at Wakefield, Massachusetts, for a station stop. Lit marker lamps and unlit headlight are the tip-off that this train is working in "push" mode. (photo by Brian Solomon)

Among the more than 7,600 F units built for North American service, F3A built as Gulf, Mobile & Ohio 884A is a true survivor. See Chapter 2, page 38, for a photo of this locomotive in GM&O paint. Rebuilt in 1978 as MBTA No. 1151, it worked for over a decade in Boston suburban service. Later it became Metro-North No. 412, and finally it was sold to the Adirondack Scenic Railroad where it became No. 1502. As of 2004 it remained in regular service. (photos by Brian Solomon)

On December 7, 1992, recently acquired Metro-North FP10 No. 412 leads the Bridgeport-Waterbury local train at Stratford, Connecticut. At Devon, this train will diverge from the North East Corridor and travel up the single track Waterbury Branch. (photo by Brian Solomon)

Close-up of the number board and marker lamp on Metro-North FP10 No. 413. When these locomotives were rebuilt by Paducah in 1978, one change was the application of extended cylindrical marker lamps in place of traditional class lamps. (photo by Brian Solomon)

Chicago & North Western home-built nine road slugs using F B units in 1971. Three were converted from former Chicago & Great Western F3Bs, one from a Bessemer & Lake Erie F7B, and the remainder from Burlington Northern F7Bs. According to Spring 2004 *North Western Lines*, conversion work was largely performed at company shops in Huron, South Dakota, while one unit was converted at Chadron, Nebraska. These units were typically operated in conjunction with four-motor units such as GP35s and assigned to freight work on the railroad's western lines in Nebraska, Wyoming, and South Dakota. They were numbered BU30 to BU39. The extent of external changes varied as carbodies were reworked reflecting the substantial internal changes. On some units, air-intake vents on the sides were filled in, while porthole windows and roof fans were removed. Some of the former F7Bs retained their stainless-steel air-intake grills, and porthole windows were merely painted over. The units retained their essential running gear, including Blomberg trucks, traction motors, etc.

Commuter Fs

Several modern commuter-rail agencies have employed heavily modified F units in various capacities including as locomotives, power packs to provide head-end power, and as cab-control cars on push-pull sets. In most instances these F units have had substantial internal and external modification.

In 1978 and 1979, Illinois Central Gulf's Paducah, Kentucky, shops rebuilt 19 former Gulf, Mobile & Ohio F units (18 F3s and one F7) for commuter service on the Boston-based Massachusetts Bay Transportation Authority (MBTA). MBTA provides heavy-rail suburban passenger service on Boston & Maine, Amtrak, and Conrail (now CSX) routes radiating out

Metro-North No. 413 rests at Danbury, Connecticut, on November 16, 1992. (photo by Brian Solomon)

Side-view detail shows the air-intake vents on Adirondack Scenic Railroad No. 1502, formerly Metro North 412. (photo by Brian Solomon)

RECYCLED F UNITS

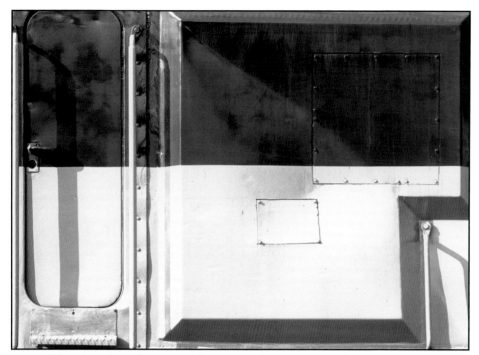

One of the identifying features of the FP10 is the extended side panels at the back of the locomotive needed to accommodate the HEP generator, equipment not found on traditional F units. (photo by Brian Solomon)

from Boston's North and South Stations. Locomotives 1100 to 1114 were rebuilt with HEP generators, while 1150 to 1153 were rebuilt with steam heat with provision for HEP, and later converted to HEP. The locomotives retained essential equipment such as the 16-567B engines, trucks, frames, and structure, but involved substantial modification and new equipment. The traditional electrical system was replaced with modern solid-state electronics. Air flow was updated with a system designed by Paducah Shops. Externally, very little of the original locomotives remained. Side panels were replaced, porthole windows removed, and side air-intake vents were covered by small filter grills rather than by integral side panel louvers. At the back of the locomotive, the carbody was adjusted to make room for the HEP generator. This modification is clearly

The fireman's side view from Adirondack Scenic Railroad No. 1502 as the locomotive leads an excursion train at Utica, New York, heading toward Thendara. One advantage of carbody locomotives such as the F unit was a superior forward view. (photo by Brian Solomon)

Long Island Rail Road F9AM 622 at Port Jefferson, New York, on March 9, 1996. LIRR designates this as an "Auxiliary Power Control Unit." Rebuilt from B&O F7A 265, later B&O 4524, this unit served as a control cab and HEP power unit for a push-pull suburban passenger set. (photo by Patrick Yough)

MARC is a Baltimore and Washington, D.C., suburban passenger operator. It used rebuilt former Baltimore & Ohio F7As in commuter service. MARC No. 81 pauses at Point of Rocks, Maryland, on April 20, 1989. (photo by Doug Eisele)

evident by the modest extension beyond the traditional carbody paneling. Nose sections were modified with a modern marker lamp design replacing the traditional classification lights above the number boards. Above the anticlimber, modern plugs for multiple unit and HEP cables altered the smooth, curved appearance of the traditional "bull dog" nose. In the cab, modern control stands replaced traditional F-unit stands. Additional roof modifications reflect changes to the airflow and engine exhaust systems.

MBTA used the FP10s intensively for a decade and then began setting the units aside in the late 1980s. It considered rebuilding them a second time, but according to issue 89 of *Extra 2200 South*, this option was deemed too expensive. The magazine also states that one of the problems with the FP10s was a reliability issue with HEP generators. In their last years of service on MBTA, the FP10s primarily hauled trains of old Budd-built RDCs for which HEP was unnecessary. During the early 1990s, MBTA sold off its FP10s. Among the buyers was Metro-North, which picked up four units — Nos. 1109, 1113, 1151, and 1152. They were repainted and renumbered Metro-North 410 to 413. These were assigned to passenger shuttles on Connecticut branches and on local trains on both Harlem and Hudson lines, primarily outside of electrified territory. Cape Cod Railroad acquired 1100, 1101, and 1114.

Morrison-Knudsen rebuilt five former Baltimore & Ohio F7As for the Maryland Department of Transportation in 1981. According to issue 72 of *Extra 2200 South*, the locomotives were upgraded to F9 standards rated at 1,800 hp and equipped with auxiliary 350 kW HEP generators. Among the external modifications were a new boxy pilot, twin sealed beam headlight, new number boards and marker lamps, new side panels, and removal of side porthole windows. Radiator and dynamic brake hatches were modified to accommodate the HEP gen-

In 1981, Morrison-Knudsen rebuilt Baltimore & Ohio F7s with HEP generators for Maryland Department of Transportation suburban services, later operated by MARC. Two MARC F9PHs idle at Washington Union Terminal on November 6, 1992. (photo by Brian Solomon)

Toronto's GO Transit used Ontario Northland and Milwaukee Road FP7s as Auxiliary Power Control Units on its suburban passenger services. Three of the APCUs lay over at Guelph Junction, Ontario, on September 24, 1988. (photo by Brian Solomon)

erator. The locomotives were numbered 7181 to 7185, former B&O 4580, 4582, 4566, 4472 and 4557, respectively. An additional unit was rebuilt for HEP but was not used as a locomotive.

Toronto's GO transit used rebuilt former Ontario Northland and Milwaukee Road FP7s as cab-control cars. Trains were operated in push-pull fashion, powered at one end by a road-switcher type locomotive, with the converted F serving as a cab and HEP power unit. In a similar fashion, Long Island Rail Road used old Alco FA cabs and Electro-Motive F units as push-pull control cab/HEP power units.

Among the external changes made by Morrison-Knudsen to these commuter Fs were the introduction of twin-sealed beam headlight, bulk pilots, sealing up of the front nose door, and replacement of class lamps with boxy marker lamps. (photo by Brian Solomon)